Residential Property Appraisal

Residential Property Appraisal, Volumes 1 and 2 are essential handbooks not only for students studying surveying but also for surveyors and others involved in the appraisal of residential property.

Volume 1 has been updated and covers the valuation process as it relates to residential properties, particularly when valuation is undertaken for secured lending purposes. It addresses the basic skills required, the risks posed in a valuation, the key drivers of value, emerging issues that impact valuation and the key legal and RICS Regulatory considerations that a valuer needs to understand.

Volume 2 of the book goes on to address the inspection and survey of residential properties, covering new technology, modern methods of construction, problem plants and pests, damp in new builds, and modern building services. New challenges for the surveyor to consider include the health and well-being of building occupants, the Party Wall etc. Act 1996 and schedules of condition, energy and building performance, and owner-occupied and tenanted properties.

An essential book for students studying to enter the residential survey and valuation profession and for existing practitioners who wish to improve their knowledge of industry practices.

Chris Rispin is a former practising surveyor focussing on the valuation and survey of residential property in South Yorkshire. His national responsibility with corporate organisations allowed him to develop guidance and training programmes for Surveyors. Phil and Chris created BlueBox Partners as an independent training organisation supporting surveyors in developing their technical skills. He is now semi-retired, assessing students for the Sava Diploma in Residential Surveying and Valuation.

Fiona Haggett is Head of Valuation at Barclays UK, responsible for setting lending policy, advising the Bank, overseeing the technical activities of the Bank's valuation partners and dealing with all valuation-related concerns. She is chair of the Valuer/ Lender Liaison group and a steering group member of ReSLA, which is looking to drive change and standards within the industry. She is also a director of BlueBox partners, and was previously UK Valuation Director at RICS, overseeing industry initiatives and the development of standards and guidance across all valuation disciplines.

Carrie de Silva was a Principal Lecturer in law and taxation, teaching prospective rural chartered surveyors at Harper Adams University, Shropshire, for over 20 years. She has an interest in the legal frameworks of owning and managing rural land, professional negligence and civil liability, and criminal liability for personal injury or death, particularly in the agricultural, forestry, equestrian and property sectors. Carrie co-edited and rewrote *Galbraith's Construction and Estate Management Law* (Routledge) and also works on the history of professionals in the property sector.

Phil Parnham is a Building Surveyor and a Member of the RICS with 40 years' working experience in the surveying profession. Initially Phil worked for public housing authorities before moving to Sheffield Hallam University to teach on their building surveying courses. Phil moved to BlueBox Partners in 2006 and has been involved in a wide range of training activities. He has also written a number of technical books and professional guidance documents for RICS.

Larry Russen is a Chartered Building Surveyor with over 40 years' practical experience in residential, industrial and commercial property. He is a Chartered Building Engineer and party wall surveyor and established Russen & Turner Chartered Surveyors in King's Lynn in 1981. Over the last 20+ years he has enjoyed training thousands of surveyors, engineers and energy assessors with RICS, Sava and BlueBox partners. His mantra for any surveyor, aspiring or practising, is 'it's all about your attitude and attention to detail'.

Residential Property Appraisal

Volume 1 – Valuation and Law

Second Edition

Chris Rispin, Fiona Haggett, Carrie de Silva, Phil Parnham and Larry Russen

Routledge
Taylor & Francis Group

LONDON AND NEW YORK

Cover image: © Allan Baxter/Getty Images

Second edition published 2022
by Routledge
2 Park Square, Milton Park, Abingdon, Oxon OX14 4RN

and by Routledge
605 Third Avenue, New York, NY 10158

Routledge is an imprint of the Taylor & Francis Group, an informa business

First edition published by Routledge 2000

British Library Cataloguing-in-Publication Data
A catalogue record for this book is available from the British Library

Library of Congress Cataloging-in-Publication Data
Names: Rispin, Chris, author. | Parnham, Phil. Residential property appraisal.
Title: Residential property appraisal: valuation and law / Chris Rispin [and four others].
Description: Second edition. | Abingdon, Oxon; New York, NY: Routledge, 2022. |
Original edition published in 2001 with Phil Parnham as the first named author. |
Includes bibliographical references and index.
Identifiers: LCCN 2021029535 (print) | LCCN 2021029536 (ebook) |
ISBN 9780367419615 (hardback) | ISBN 9780367419622 (paperback) |
ISBN 9780367816988 (ebook)
Subjects: LCSH: Residential real estate–Valuation–Great Britain.
Classification: LCC HD1389.5.G7 P37 2022 (print) |
LCC HD1389.5.G7 (ebook) | DDC 333.33/820941–dc23
LC record available at https://lccn.loc.gov/2021029535
LC ebook record available at https://lccn.loc.gov/2021029536

ISBN: 978-0-367-41961-5 (hbk)
ISBN: 978-0-367-41962-2 (pbk)
ISBN: 978-0-367-81698-8 (ebk)

DOI: 10.1201/9780367816988

Typeset in Times New Roman
by Newgen Publishing UK

Contents

Preface

Residential Property Appraisal was originally published in 2000 and this book is an updated version of that text. For many years, Phil Parnham and I resisted updating the book mainly because of the superfast developments that were taking place, with the release of information through the internet. Nowadays, it is impossible to produce a hard copy book and expect it to remain up to date for any length of time. However, what we have noted is how all that information is used needs some structure. The core principles of valuation and surveying are sustainable, but the technology and developments surrounding those may change and the professional surveyor must keep up with those and have the research ability to do that. Therefore, what we have tried to do is identify a series of benchmarks that highlight the core principles. It will be for the surveyor to consider when something changes and demonstrate to the customer how this influences the valuation and the repair or maintenance of the property in the report produced.

To reflect the need to keep up with all the changes the original authors felt it important to take on board some specialists, so in respect of valuation, Fiona Haggett has joined the team. In respect of construction and building pathology, Larry Russen has joined the team, and to demonstrate the significance of the law, Carrie de Silva has joined us.

It is now 21 years since *Residential Property Appraisal* was born, but instead of coming of age and a mere move into adulthood, the book has produced two offspring. In order to demonstrate the principles, we do need to reflect the amount of information that is now available to assist in completing a valuation or a survey. This was going to produce a huge book, so we decided to issue it in two volumes. This volume will focus on the valuation of residential property and the law supporting it. Volume 2 will focus on the construction and building pathology of residential property. However, that volume will also have law and valuation references to demonstrate how those matters that may materially affect value. An important aspect of Volume 2 is that it is not just about defects but also environmental and health and safety considerations, which now play a very important part in the work of property professionals.

I would also like to add that this book is dedicated posthumously to some of those who led various initiatives at the RICS who were not, possibly, recognised as much as they should have been for the work they did for the profession: David Dalby, Barry Hall and Graham Ellis. I worked closely with all of them, and many more who gave up their time to debate what was the best for the profession.

Most of all, recognition is due to my co-authors and not forgetting Phil Parnham and Larry Russen. The fact the book is split in two is not an indication that there has been any parting of the ways, it is merely a reflection on how residential surveying and valuation has grown in significance and emphasis to the housing market.

Chris Rispin

Notes on the text

Geographical extent

The principles of valuation that we have outlined apply equally in England, Scotland, Wales and Northern Ireland. However, the law in a number of areas can diverge across the four nations. For example, land tenure and the process of conveyance are different in Scotland and planning law is a devolved matter across all four nations (although there are many areas of commonality). Further information should be sought on legal matters if you are working outside England. In addition, you might find the process of valuation is slightly different.

Case citation

Note that, for simplicity, all case dates have been shown with round brackets (). You will see in other sources that square brackets [] are sometimes used. This is where the date is required to find the correct law report, i.e. where reports are by year rather than by sequential volume. With the ease of internet searching, no citations have been given but when citing cases formally, for example in academic or professional writing, you will need to ascertain the correct form for any particular case. Full details of the citation of cases, and all other legal materials, may be found in *OSCOLA, The Oxford University Standard for the Citation of Legal Authorities* (2012).

Acknowledgements

We would like to thank all the team at BlueBox Partners for their continued contributions to the technical discussions that ensure we have a pro-active view on our industry. We would also like to thank the team at Sava who had the confidence to produce the learning tool that is the Diploma in Residential Surveying and Valuation as a means for developing the future residential surveyors and valuers.

1 Introduction

1.1 Context

In this first part of the twenty-first century, the pace of change in valuation has been incredible. We have experienced a property market crash, the likes of which we had never seen before, and the rise of technology, which is now influencing all aspects of a surveyor's life. The surveying profession was slow to respond to these technological changes but has now become aware of the threats and opportunities that technological progress is bringing.

Various catalysts continue to influence the direction of the profession. In respect of residential property, in particular, they include:

- continuing government intervention in the property market, as it seeks to make home buying easier and more affordable;
- a drive to address housing shortages through increased house building targets and new construction techniques;
- various changes to the regulatory and professional support provided by the Royal Institution of Chartered Surveyors (RICS) and the potential rise of competing trade bodies;
- the increasing pace of technological advances that has seen Big Data playing a greater role in the industry and the large software suppliers becoming more influential players in the market;
- ever-increasing consumer standards driving a more litigious industry, where the profession needs to respond with a quality service or succumb to rising insurance costs.

This book analyses the core skills that a surveyor needs to carry out appraisals of residential properties in a fast-changing world. It is not targeted exclusively at experienced surveyors but is also intended to inform and encourage the student/trainee (from whatever background), to develop the skills associated with inspecting and valuing residential properties.

Looking back over the last two decades, it is clear to us why there is a need for this book. As we emerged from the financial crisis and property crash of the early 2000s, and dealt with the issues that this raised, the profession had to re-evaluate its practices and evolve to ensure we would never again be exposed in the way we found ourselves after the crash. Surveyors are adapting to new requirements and are generally eager to learn from past mistakes.

DOI: 10.1201/9780367816988-1

Attendance at continuing professional development (CPD) events and courses has increased, reflecting the need to maintain knowledge and understanding of technical and procedural changes. There is a new drive to bring fresh blood into the profession to replace the professionals lost in the crash and the ageing workforce who are close to retirement.

The drive to increase technical skills has come not only from a desire to keep pace with change, but also for other reasons, such as:

- The focus of the knowledge and experience of many professionally qualified surveyors has been closely associated with the valuation of property, and not just its physical state. On the other hand, the building surveyor or other similar qualified professional's focus is on the property. Few courses have catered for the equal combination of skills.
- Many educational courses lack a sound technical grounding. The professional institutions often call for broader, more flexible surveyors armed with business and commercial skills. As a consequence, a number of traditional disciplines have disappeared from the curricula. Building studies and building defects are two such subjects for the residential practitioner.
- Technical advice and guidance that are currently published have a broad audience. A lot of the literature is specifically written for those who carry out in-depth surveys and investigations of residential property. It is often difficult to identify which part of this advice is best suited to the professional practice of residential valuation and appraisal.

Court cases in the 1980s fashioned many of the surveyors' working practices and RICS Guidance followed to clarify what was required. The court cases have not ceased, although many now refine existing thinking or re-emphasise certain practices, where perhaps Guidance has not been as strong as it could be, or some surveyors just needed a reminder of how crucial the core skills really are. To try and meet this demand and plug the skill gaps, the authors have had to re-write, interpret or specially create training materials that match the professional role of the participant surveyors. There is a weakness in any written material as it cannot be updated speedily, therefore we have tried to develop a series of benchmark measures which the practitioner can use as a basis for good practice, but also develop techniques to adjust to the changes. Regular CPD is, however, still required to develop that process.

In addition, there are two other sources of change that have emerged in recent years:

- The structural changes within many lenders have resulted in different business priorities. New technology has revealed opportunities not previously available and exposed different working practices abroad that could be imported for use in this country. The structure of the residential market in countries like the USA, Australia and New Zealand is much less complex and uncertain. As the global economy has a greater influence on the domestic market and more financial institutions operate internationally, pressure will be applied to simplify the process.
- Customer expectations are changing. The typical 'customer' is becoming far more sophisticated. They are better educated and used to more customer-focused service

in the other products they purchase. Evidence from consumer surveys, media sources and pressure groups suggest that the standard of service that many surveyors provide continues to fall well below current expectations.

This book sets out to meet these challenges.

What is needed is a publication that has a clear technical focus directly related to the professional role of its readership. Because this role is closely associated with value, then this part of the book must acknowledge the actual process of valuation. Because the commercial world is never static, all this must be set against a background of change and increasing expectations from the people who matter – the fee payers!

1.2 Objectives of the book

Based on this contextual review, the aims of this book are as follows:

- to provide surveyors with sufficient practical and detailed information so that those matters that materially affect the value of residential dwellings can be appropriately assessed. In this case 'appropriately' would equate to the standard currently expected of the Mortgage Valuation, although we also cover some of the other forms of residential valuation at a high level;
- to help surveyors further develop the skills of report writing and communicating the results of these assessments to their customers;
- to provide an overview of the valuation process with a particular emphasis on how the condition of a property and a multitude of other matters such as environmental impacts affect its value;
- to highlight the changing nature of the residential property appraisal process and initially identify some of the techniques and mechanisms that may help surveyors adapt to this changing environment;
- to reflect the important role that land and property law plays in assessing the value of an interest in property, and this will feature significantly.

1.3 Definitions

Clearer definitions of the principal terms employed in this book may be useful:

Customer or client In this book, these two terms are used to describe the end user of any survey or inspection report and this person will usually be a private individual.

Residential appraisal An act or process of estimating the value, worth or quality of a residential property.

Residential property Any property that is used as, or is suitable for use as, a residence. This book is restricted to single domestic dwellings owned by individuals, rather than a corporate entity.

Surveyors This generic term has been used to describe chartered surveyors or other suitably qualified practitioners.

1.4 Who this book is for

This book has been written for a broad range of surveyors whose primary interest is the valuation appraisal of residential property. The book makes certain assumptions:

- the reader has already satisfied, or is close to satisfying, the academic requirements of their chosen professional institution. This would have included a course of study that introduced participants into how dwellings are designed and constructed and the principal agents responsible for the deterioration of the building fabric.
- the reader has had some professional experience of assessing a range of different properties.

In terms of qualifications and level of experience, this book should be suitable for:

- student general practice surveyors in the later stages of their academic course or on the sandwich placement or year-out stage;
- building surveyors who are looking for an introduction to the assessment of residential properties;
- students on courses of a more specialised technical nature who need an understanding of how to carry out an appraisal of residential property;
- surveyors who are working towards their professional assessment and need to refer to written guidance and technical information on a regular basis;
- those more experienced surveyors who may be changing their professional emphasis towards residential valuation;
- qualified and experienced surveyors who need to carry around a source of reference so they can refer to standard guidance when novel or unusual situations are encountered.

1.5 The philosophy of the book

The guidance contained in this publication aims to be challenging to the reader in two ways:

- to outline processes and techniques that may potentially take the surveyor beyond the parameters of current standard valuations. This will enable surveyors to more effectively provide those services and better cope with any changes to standard practice in the future.
- to engage with the surveying process and positively advise clients about the suitability of the price for their potential new home.

This book is not a guide to any particular standard form of valuation promoted by any particular corporate body or lender. For something more specific, then the reader should refer to the publisher of the particular product.

1.6 Contents of the book

The book is divided into two volumes:

Volume 1 starts with an overview of the valuation process itself. Particular emphasis is placed on the role of the condition of the property in determining value. Volume I then moves on to cover the principal elements of law as they affect the property professional.

Volume 2 covers all the defects normally associated with residential property and outlines a strategy to resolve these problems. In addition, it gives practical guidance relating to good practice in report writing. This includes a number of case studies to illustrate both good and bad practices.

2   Overview of the valuation of residential property

2.1 Introduction

Valuation is a key part of the appraisal process for residential property and this chapter covers the most common valuation approaches used in this sector of the market, in particular, the **comparable method**, supported by the use of benchmarking tools. The adoption of a step-by-step scientific approach in applying comparable evidence removes some of the subjectivity from the process, resulting in more certainty and greater accuracy. Whilst it will never be possible to produce a black-and-white solution to the valuation of residential housing, it is hoped that a conclusion reached using a methodical and documented approach and which is based on facts supported by market knowledge will narrow the 'grey' areas and give greater clarity to the final outcome.

2.2 Value

2.2.1 What is value?

Before we get too involved in those criteria that influence the value of a property, we should briefly outline how the valuation of a property takes place and then we can build on that foundation.

The value of one specific house is derived by looking at similar houses that have been sold recently that are nearby. So ideally the comparable properties look similar in that, if the subject is a detached house, then the comparables should be detached houses with the same number of bedrooms, be a similar size and be in a similar condition. They should also have the same facilities such as heating and car parking provision, together with similar sized gardens and they will have been on the open market for a period of time with the sale having completed.

That is the ideal scenario, but this is often not the case. It is in situations where there are no similar comparables, or they have not all sold recently, or there is something about the legal interest that differs and possibly they are not in the same sort of condition, that the skill of the valuer comes to the fore. This chapter will look at those attributes that make the difference and take the reader through how to deal with them.

In the case of *Roberts v J Hampson & Co.* (1989), Ian Kennedy described an important aspect of a valuation for secured lending: 'It is a valuation and not a survey,

DOI: 10.1201/9780367816988-2

but any valuation is necessarily governed by condition.' This case related to a building society valuation, but the principles apply just as much to any valuation for whatever purposes. Judicial precedent has dictated many of the principles that apply to a mortgage valuation and such is the influence of those in the market that many other forms of valuation (including valuations for Inheritance Tax or expert witness purposes) use principles that have been developed in the courts over many years.

See Chapter 4 for an overview of the operation of case law and the status of cases heard in different courts.

When this case was recorded, the average mix-adjusted house price for the UK (according to the Office for National Statistics) was £60,000. In 2021, when this book was published, the average UK house price was in upwards of £300,000 and this has put a different perspective on value. In the 1990s, condition continued to be a significant feature in assessing value, but as standards changed and house prices escalated, condition became less significant and this was overtaken by other features, such as estate charges, lease length and other legal issues.

! **Benchmark: House Price Index**

This is our first benchmark that recognises a change – the **House Price Index**. As prices rise or fall, then how does the current figure compare to the last recorded price based on a property price index? Is the local market supported by the national or regional indices or are local factors more significant?

The importance of buying a house in the condition expected by the purchaser will be considered in this chapter relative to value, but we will also look at a number of other factors that now need to be included in the definition of elements that materially affect value.

Since the original edition of this book was published, the amount of data that is freely available to the surveyors and their clients has changed dramatically. All are now much better informed, there can be fewer assumptions and, if there *are* assumptions, the degree of uncertainty can be reduced. This should make for more accurate and reliable valuations.

Before we consider anything else, it must be made clear that anyone undertaking a valuation of property for any credible purpose should have working knowledge of the current RICS *Global Valuation Standards*, otherwise known as the 'Red Book', as this is a key reference manual. The UK Supplement to the Global Red Book is an additional document that supports the RICS published standards by applying the global principles to the UK market. This is of great help to the inexperienced valuer, as it gives an insight into the process necessary to achieve a competent appraisal of residential property. At the time of writing, in the UK there were no other valuation standards available, but that may change over time.

> **! Benchmark: Standards and the RICS Red Book**
>
> This benchmark is **Standards** and specifically the **RICS Red Book**. What standard was applicable when the valuation was undertaken? As those standards change, then what impact does that have on the valuation? At the time of writing, fire risk in high rise buildings has become a key issue and standards have changed to reflect that. But for valuations undertaken prior to this change, then could they have reflected that external wall cladding was in some cases flammable and thus a known risk?

A valuation usually takes place when a property changes ownership or there is a change in circumstances, but the basis and form of advice may vary. This depends upon whether the parties to the transaction are buyers, sellers, estate agents, conveyancers, financial intermediaries or lending institutions. However, the underlying theme for all valuations is establishing whether the property is a 'good deal' for the stated purpose. For example, a purchaser will want to know if it is good value for money, while a vendor, on the other hand, wants to know that they are not underselling the property. From a lender's perspective, it is important for them to know that the property is good collateral for the loan and any potential risks associated with achieving a resale on possible repossession. This publication concentrates on the form of advice that those parties may need by specific reference to 'factors revealed during an inspection that are likely to materially affect its value' (The Valuer's role and remit, UK Valuation Practice Guidance Applications [VPGA] 11.1.7). It is for the practitioner to establish the precise needs of the client and to provide the advice in the form requested and/or required by the RICS Red Book, plus associated Guidance issued from time to time by the RICS, such as the Valuation of New Build Homes.

In its broadest sense, the value in a property is made up of everything that is included in the interest; both the land and the structure on it. However, it is never that simple. Property is, as a product, heterogeneous – no two properties are alike. Likewise, the purchasers of a property are quite often dissimilar in the features they are looking for. This brings with it a huge range of complex issues that the valuer has to understand and unpick. The personal nature of domestic likes and dislikes is what makes the valuation of residential property so dissimilar to the hard-headed and logical business decisions that govern the value of commercial property.

To begin with, it will be useful to briefly review the meaning and interpretation of key words by reference to both the dictionary and the RICS Red Book. To assist in this process, the respective definitions of a number of terms are compared (see Table 2.1).

We have provided some key definitions and the property-related definition as opposed to the dictionary definition because the inferences can, in some cases, be slightly different. Price, although no longer defined, is used throughout the Red Book with various tags, such as 'best', 'fair', etc. However, it is clearly stated that 'price' does not necessarily equate to value as attributed by the valuer, depending upon the conditions that have to be applied. So, as can be seen above, the market value is constrained by a number of factors, some of which may not have applied in the transaction, but it is the valuer's role to ensure comparables are selected that mimic those requirements.

Table 2.1 Comparison of key valuation terms

Definition	RICS definition	Dictionary
Value	Market Value – The estimated amount for which an asset or liability should exchange on the valuation date between a willing buyer and a willing seller in an arm's length transaction, after proper marketing and where the parties had each acted knowledgeably, prudently and without compulsion (see IVS 104 para. 30.1).	The **value** of something such as a quality, attitude, or method is its importance or usefulness. If you place a particular **value** on something, that is the importance or usefulness you think it has (Collins, 2014).
Worth	Investment value (worth) is defined in IVS 104 para. 60.1 as: 'the value of an asset to a particular owner or prospective owner for individual investment or operational objectives.' As the definition implies, and in contrast to market value, this basis of value does not envisage a hypothetical transaction but is a measure of the value of the benefits of ownership to the current owner or to a prospective owner, recognising that these may differ from those of a typical market participant. It is often used to measure performance of an asset against an owner's own investment criteria.	Someone's **worth** is the value, usefulness, or importance that they are considered to have (Collins, 2014). Could equally be applied to 'something'.
Appraisal	No longer defined in the Red Book but used frequently in association with 'valuation'. Previously defined as the valuation process, assimilation of information sufficient to create an audit trail.	**Appraisal** is the official or formal assessment of the strengths and weaknesses of someone or something. Appraisal often involves observation or some kind of testing (Collins, 2014).
Price	Price is no longer defined in the Red Book, although it used to be: 'The amount of money someone would have to pay in the open market to replace something they have been deprived of.' One definition can be found under the Charities Act section of the Red Book, referring to the case of *Duke of Buccleuch v IRC* (1967) (but there are others): 'the price that the property might reasonably be expected to fetch was defined as the gross sale price for the property without deducting any selling costs.'	The **price** of something is the amount of money that you have to pay in order to buy it (Collins 2014).

When undertaking a valuation, it is critical, therefore, to establish what interpretation of price is required to achieve the value. This is different for an Inheritance Tax valuation and also one under the Charities Act.

The linkage between all these terminologies is that the surveyor must go through a recognised process (**the appraisal**) to accumulate information that is sufficient to establish the property's relative quality (**value**) and its specific range of benefits to the client (**worth**). This is so a monetary figure can be attributed (**price**) which could be considered as compensation should the client be deprived of those benefits.

2.2.2 The elements of value

We all know that the biggest influencer on the value of any property is 'location, location, location'. Historically, a higher demand has developed for properties in the South East of the country and thus prices in these areas are higher than for similar units in the north of the country.

However, the micro elements associated with a location will also drive the desirability and, consequently, the value of the property. Factors such as orientation, views, proximity of infrastructure, rights of way, restrictive covenants, noise and pollution will all serve to attract or deter a purchaser.

This all sounds very simple, and in some cases it is. However, the more unusual or complex a property is, the more difficult it becomes to identify all of the elements that make up the property and their relative impacts on the end value.

If we break down these elements of value, they will look something like the diagram in Figure 2.1.

From this we can see that the base value is largely determined by location, but on top of this we have to start looking at other factors. Beneath the ground, elements such as soil type (chances of subsidence occurring), flooding, past land uses (e.g. waste tips, filling stations, mining) and naturally occurring gases, could all affect the buying decision.

In addition to this, the property attributes (those tangible elements in the structure of the building), will all play their part. This may be the number of bedrooms, floor area, room layout, parking and garages. These elements are easy to identify and compare when considering one property in comparison to another. However, residential property valuation becomes more complex when you start introducing intangible elements, or those factors that can be considered as personal preferences and which are not so easily identified. These will be things such as unusual property features, a particular style of kitchen, the demographics of the area, orientation of the garden, views, and the like. It is in valuing these unique properties that the valuer faces their biggest challenges and has the greatest opportunity to prove their worth.

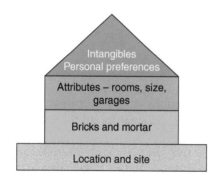

Figure 2.1 The elements of value

> **! Benchmark: Core attributes**
>
> The benchmark here is the **core attributes** of the property – its size, number of bedrooms, detachment style and provision for parking. These are the tangible items that are the basis for comparison so that the valuer is comparing 'apples with apples'. They will vary depending on the type and style of property.

In addition, in making the valuation judgement, competent professionals also have to know their market. Not only in terms of the geographical area, but also in relation to the nature of a likely purchaser for the type of property in question. The housing market is made up of a number of different groups, at all stages in their lives. The first-time buyer will have very different buying priorities to the family unit and the decision-making process changes again as people age, downsize and move into retirement. Whilst location always remains critical, space is at a premium for growing families and usually declines in importance in later life. What does start to creep in as important at this point is accessibility and comfort. One element of comfort is the energy efficiency of a property and, with the government seeking to drive the sustainability agenda, we are likely to see the environmental credentials of a property becoming more influential as a value driver in the market.

The issues and complexity come from the fact that current data does not record all of the elements defined by Figure 2.1. It is relatively easy to compare a property of a similar size with the same number of bedrooms and of the same style in the same area, provided there have been numerous sales recently. However, where property is of a different style and in a slightly different location, then personal preferences have to be accounted for and this is where the risk increases, because it is a more opinion-related assessment based on intangible elements, rather than tangible facts.

The impact of government activity is the final consideration in this section. There is an increasing volume of legislation influencing the market and which is distorting the prices in certain sectors or areas. These impactors include:

- Stamp Duty Land Tax (second homes, thresholds, etc.)
- Taxation of landlords
- Minimum energy-efficiency standards
- Licensing of rental properties
- Affordable housing initiatives
- Section 106 requirements (referring to the Town and Country Planning Act 1990, as amended, see Section 10.5.4).

Today's valuer needs to have a comprehensive understanding of these value drivers and also needs to keep one eye on upcoming changes that could have an impact.

The conclusion to be reached from all of this is that the valuation of some properties that are mainly composed of standard, tangible, elements can often be considered to be straightforward and a low-risk exercise. It is for this reason we have seen the rise of automated valuation models (AVMs – see Section 10.8.5). However, as the number of intangible elements that relate to a property rise, the greater the complexity of the valuation and the higher the risks associated with it. It is always worth bearing this in mind when approaching a valuation and focusing your attention accordingly.

2.2.3 An interpretation of the extent and nature of the interest

There are a whole series of specific attributes attached to a house that can best be described as those issues that affect the client's rights to use the property. These may be either to the advantage or disadvantage of the client and consequently will influence the value of the property generally, and the worth to the client. We have set them down here for context but for more information on the legal aspects, see Chapter 5. In addition to the legal elements in this book, there are numerous books on property law that will give further detail and legal background to these issues.

The main property law issues of impact to valuation are:

> **Tenure**: the legal right to use the land. This can be freehold or leasehold in England, Wales and Northern Ireland. (This book does not consider Scottish law. Some of the valuation principles will apply, but land and tenancy law differ significantly.)
>
> The property may be subject to certain forms of ground rent, service charge or chief rent. This is an additional cost on the right to use the property. The amount and the conditions surrounding the charge will vary and some will be significant in arriving at the value.
>
> **Planning issues**: rights under the planning legislation convey the parameters under which the owners can legitimately use or develop the land and its property, for example, residential or commercial occupation. The right to convert land used for agricultural purposes to residential would usually add significant value as the return is greater. On the other hand, a restriction on use may adversely affect value (see Chapter 5 for more details).
>
> **Building Regulations:** Building Regulations govern the actual structural design of the property especially in ensuring that the construction is safe, whether for new construction or adaptations of the existing structure.
>
> **Rights of light, of way and other easements or covenants:** these are legal interests or contractual rights which might be found in the deeds or Land Registry documentation to the original purchase. They can be modified over time and they can be restrictive. Examples could include:
> - where a developer and the planning authority agree to an open plan estate that restricts the development of hedges or walls;
> - where rights were given allowing a person access to a piece of land over someone else's property. The initial grant might have been given in exchange for some form of consideration (fee) to reflect the adverse impact of the access. Subsequent sales will need to reflect that impact.

The overall impact of these restrictions may be to the benefit of an area as a whole. For example, a right of way gives access to many, but may impact the principal occupier of the land. The benefit may need to be reflected in valuing neighbouring property.

It is important to determine whether these rights have simply been agreed between two parties, and cease when one of those parties sells their property or dies *or* whether they are proprietary rights which bind all future owners and occupiers (see Chapter 5).

Benchmark: Legal rights

The benchmark here is the **legal rights** impacting upon a property. The assumption will be that they are normal and not onerous, so anything that deviates away from that benchmark may need to be reflected in the value.

Environmental considerations – This covers a wide range of factors varying from climatic conditions to previous or current contamination of the site, such as the presence of radon gas. Again, the scenarios can be either beneficial or adverse, in that they can enhance or detract from the value. Climatic conditions will feature in various parts of the condition section and reference is made to contamination because of the need to ensure correct site treatment to prevent escape of the contaminant (see Volume 2). The latter aspect poses additional costs in the original construction that may well not be capable of being reclaimed through the sale price, when compared with other developments without the contamination. In these cases, the developer may choose not to build property until profit margins allow.

The legal features described above may not be apparent from a site inspection of the single dwelling. Reliance is often placed on the conveyancer to unearth any matters that may adversely affect price or value. The surveyor should be aware that the conveyancer is depending on them to act as the 'eyes and ears' to identify a 'trail of suspicion'.

Consider the following example. There are many large blocks of flats in Greater London, some of which have leisure complexes. The right to use the complex can be included within the deeds, but in some cases it is not. Some leases make an additional charge for this benefit. The conveyancer will be depending on the surveyor to identify whether there is a leisure complex. This sounds straightforward. But because a lease can extend over more than one block and can include hundreds of flats, the task of locating a fitness gym hidden away in a basement might not be that easy. Take, for example, an instance where a purchaser knows about a leisure complex and assumes that the right to use it is included in the purchase. However, in the event, the complex did not appear in the deeds and the surveyor did not mention it. The potential for a legal challenge by a disappointed purchaser would be a distinct possibility. As the customer rarely talks to the surveyor and the surveyor rarely talks to the conveyancer, such a breakdown in communication is almost inevitable occasionally.

2.3 Equipment

A key requirement for any valuer or surveyor is having the correct equipment. Not having the proper equipment on a survey is no defence against an action of negligence. The list of surveying equipment will vary depending on personal preferences, the organisation the surveyor works for and the sort of work to be tackled. The following is a list of equipment used generally by a surveyor. The list has been split between what is most likely to be required for a valuation and that required for a more detailed inspection, with a generic section on health and safety items.

2.3.1 Valuation equipment

A typical list for a valuation could include:

- **measuring tape**

a laser measure, 5–7.5 metre retractable steel tape is suitable for most jobs but occasionally a 20–30 metre fabric tape will be needed;

- **electronic moisture meter** (with spare batteries)

this should be calibrated every time it is used;

- **spirit level**

a small-hand held level is useful for checking the levels on floors, the verticality of walls, as a guide to whether further investigations may be required;

- **surveyors' ladder**

four sections, others that retract or fold are useful but check whether they are suitable – they have more moving parts and so, for some people, they can be difficult to operate. The length of the ladder needs to be sufficient to undertake a head and shoulders inspection of the roof space. However, for surveys, the ladder needs to be of sufficient length to safely access a 3-m-high platform, so that may be the main consideration when selecting one.

- powerful inspection **torch** (and spare batteries)
- **clipboard and paper**

for notes and sketches, possibly more as a backup for an electronic device;

- suitable **camera** (with flash)

most modern phones can provide good quality digital photographs, but some people may prefer a stand-alone camera.

2.3.2 Additional equipment for a more detailed inspection

- a plumb bob and line – a quick and effective way of checking the verticality of walls;
- camera pole, for out-of-the-way places;
- binoculars (× 8–10 magnification).

2.3.3 Equipment for simple 'opening up'

- a robust claw hammer;
- a large flat-head screwdriver;

- a 'wrecking' or crowbar (450 mm long);
- bolster or cold chisel;
- two large and two small drainage inspection keys;
- a bradawl or other suitable probe.

NB These are most unlikely to be a requirement for a mortgage valuation, unless you find yourself locked in somewhere.

2.3.4 Health and safety equipment

- personal attack alarm and spare battery;
- mobile phone;
- first aid kit;
- safety helmet;
- face mask with disposable filters for loft inspections, etc. These must be a suitable specification for the dust and fibres that can be expected.
- safety goggles;
- pair of protective gloves for lifting inspection chamber covers, etc. Many organisations are now recommending disposable gloves because once the gloves have been used, if they are put back into the survey tool kit, they can contaminate other tools.
- disposable rubber gloves for dirty or unhealthy locations;
- appropriate steel toe-capped wellingtons or other suitable safety footwear.
- plastic overshoes.

This may seem a lot of equipment but all of it could be used on a typical valuation. It does not need to be carried around room to room, but it should be available if required.

3   The economics of property valuation

3.1 Introduction

It would be remiss to produce a text on valuation without taking a look at the impact of wider economic factors on the housing market, and indeed the valuation process. To give an idea of the importance of the sector, in 2018, the value of the housing stock in the UK stood at over 50 per cent of the UK's net worth, with the figure at approximately £7.3 trillion (Savills' blog, 4 February 2019). A quarter of that value is accounted for by London property alone. In appreciating this, it is easy to see that the wider economy and the value of property are bound to influence each other and thus have, historically, resulted in a great deal of political interest and interference in the property sector.

Those students who want to understand economics in greater depth should seek specialist textbooks, of which there are many. However, in this chapter we will take a look at the fundamental principles and a brief history of the relationship between the economy and property prices.

3.2 Supply and demand

Put simply, the value of a property is determined by the basic principles of supply and demand.

In Figure 3.1, the volume of goods produced (supply) will be determined by the purchase of goods (demand). As more goods are demanded, then, in simple terms, more goods will be produced. Where the supply and demand curves cross, then this is the determinant of price, the equilibrium. The normal production of low-value items, such as chocolate bars, involves the creation of thousands if not millions of products with just as many buyers. So, the supply and demand curve will be fairly standard.

However, housing development in this country involves much lower volumes of high-value units, which take time to produce (to build) and are located over a wide area, so the potential for the increase in supply of new properties in any one place is relatively small. Added to this is the influence on the local market of the sale of existing property, albeit in small volumes, and the demand in any one area with, usually, only a few potential buyers for each property. All of this has a significant impact on how the economics of price works.

When a property owner offers a house for sale, they are adding to the supply and, in purchasing another property, they are increasing demand and simultaneously impacting on supply in the area in which they are purchasing.

DOI: 10.1201/9780367816988-3

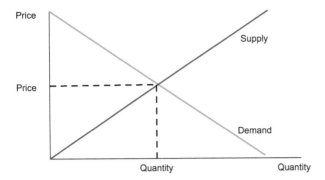

Figure 3.1 Simple graph showing the point of equilibrium or the price point based on this scenario of supply and demand

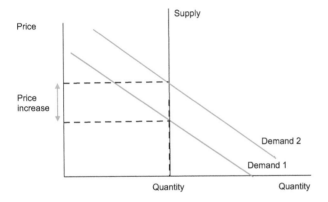

Figure 3.2 The impact on price where there is inelastic supply and differing levels of demand

The nature of housing is that the supply is said to be 'inelastic'. Land is in short supply and building takes time, so if production cannot be increased in a short time-scale, then the supply curve is depicted as nearly vertical. Put diagrammatically, the impact is illustrated in Figure 3.2.

Where the demand and supply lines cross, there is a point of equilibrium, which will determine the price. As a result, an increase in demand for one particular property (three or four potential buyers rather than one) will inevitably result in a price rise, as supply cannot be increased quickly to keep prices stable as it can with mass-produced products. In this case there is only one product, and whilst there may be another similar property nearby, this may not have exactly the features that have attracted buyers to the original house for sale.

Conversely, if there is a property in an area where there is little demand, then the seller has no option but to reduce the price to increase the market attraction, albeit there only needs to be one buyer.

Figure 3.2 would look slightly different for an area where there is a substantial amount of new building or large-scale developments (such as blocks of flats). In such instances, during the development period, the market may act just like that of a typical

What drives house prices?

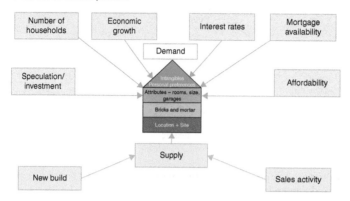

Figure 3.3 The differing drivers on house prices converted simply into supply and demand criteria.

manufactured product. This is because there are a number of units for sale and usually a number of potential buyers all at the same time.

The influencers of demand in the property market can be complex and subject to a great deal of political interference, the outcome of which is not always easy to predict (see Figure 3.3). However, the main driver will always be the ability to pay. This element is mainly driven by incomes but is also influenced by other factors such as interest rates, equity release and foreign investment. These are followed closely be physical factors such as location, size, style and user restrictions.

An increase in supply can be caused by unemployment, for example, due to major factory closures or the movement of the population out of the city due to changes in work practices (such as we saw in the COVID-19 pandemic of 2020–21) and, in many cases, the opposites of all those elements that create demand. So, a significant increase in interest rates, as we saw in the early 1990s and in the early twenty-first century, resulted in an increase in available properties due to a lack of demand for the houses on the market.

It follows that if an under-supply causes prices to rise, then an over-supply will do the reverse. An example of this has been the digital culture that has caused a decline in the demand for office space. The resulting price falls have meant that, in many cases, the opportunity cost of retaining the existing use against an alternative use have tipped the economic scales to a more profitable use as a residential conversion, helped by a relaxation of the restrictions on planning controls by the government. This change to the economics of a commodity is referred to as 'transfer earnings'.

3.2.1 Brief historical analysis of the residential property market

Having looked at the basic principles of demand and supply, it is worth taking some time to look back at the property market in the UK, particularly at major disruptions to the market, some of the reasons behind them, and how they affected property values.

Stretching back some decades, the property market 'crash' of 1990 was a severe price 'correction' coming off the back of a ready supply of cheap credit in the 1980s. The 'loadsamoney' culture at the time had encouraged people to invest in property and home ownership was increasing fast, as council houses were being sold off to existing tenants at a significant discount through the recently introduced Right to Buy legislation. At the same time, lenders relaxed lending criteria for re-mortgages and a merry-go-round of re-mortgaging to release equity from the spiralling prices of the houses gripped some homeowners.

As the 1980s drew to a close, two things happened to threaten this bonanza. The first of these was a decline in the industrial sector of the economy, which sparked a steep increase in interest rates and unemployment, impacting affordability (in the early 1990s interest rates of 15 per cent were not unusual), and the second was the decision by government to remove Mortgage Interest Relief at Source (MIRAS), a tax relief on mortgage repayments. Whilst the issue of affordability took the heat out of the market, the removal of MIRAS had a distortionary impact, simply because the government announced the changes in advance, with an effective date some months in the future. The effect of this was to cause a rush of purchasing in the market, in order to beat the deadline, followed by complete inactivity and a drastic fall in property values. As affordability declined, repossessions increased, further impacting values.

Whilst the house price correction in 1990 was severe, it was fairly short-lived and was followed by a period of economic recovery, during which time interest rates decreased, finance became more readily available and there was a consequent steady increase in house prices. This accelerated in the early 2000s, as lenders competed aggressively to attract borrowers, by introducing ever more favourable and accessible mortgages.

The crash in house prices in mid-2007, however, was due to an entirely different cause than the issues of the 1990s. Activity in the housing market in both the UK and the US had been frenzied in the run up to 2007 and no one had been entirely aware of the emerging problems in the sub-prime market (the practice of lending to borrowers with low credit ratings with higher than standard interest rates to reflect the increased risk to the lender). In 2007, sub-prime defaults in the US increased dramatically, causing lenders to fail and a crisis in confidence in the financial markets. As no one economy acts in isolation, the catastrophic impact of the crash in the US quickly spread to other countries. The impact of this loss of confidence and a long period of lax financial controls meant that activity in the economy, including the housing market, dropped almost overnight.

At this point, valuers began to 'feel the heat'. As repossessions increased and lenders began to see the crystallisation of losses, the valuers found an increasing number of claims letters (alleging negligent valuation) arriving at their offices, particularly from sub-prime lenders. These letters became known as 'confetti letters', due to the large number of them being sent out. Some offices would see 30 or 40 claims arriving at one delivery. Not all of these claims of negligent valuation were justified, indeed, significant levels of fraud were discovered with misleading quoted prices. It was possible in those days to exploit the lack of recorded data relating to housing transactions. The valuation industry learnt a sharp lesson from this period, which it is important not to forget. This is that the importance of good comparable evidence (or other appropriate methodology correctly applied), a supporting rationale, and sound record keeping can never be overstated.

The period following the 2007 crash saw increased regulation on lenders and a drop in the availability of credit, particularly on high loan to value transactions. As a result, the number of first-time buyers in the market dropped and this gap was filled by the rapidly expanding buy-to-let sector. The restriction did not begin to ease until 2012, when new products started to come through and lenders began to relax their income multiplier requirements.

Whilst the post-2012 market showed signs of recovery, the crisis in affordability continued to bite. As an illustration of the extent of this problem, it is worth looking back at the market prior to the 1980s. At this time, property values at three times median income had sustained the housing market for some years. All this changed during the house price booms of the late 1980s and early 2000s, when house prices increased disproportionately to income. Part of the reason for this was the increased availability of cheap credit, but one of the other drivers was the rapid increase in buy-to-let purchases on the back of interest-only mortgages. These and other drivers, which are not too clearly defined, resulted in house prices in London soaring to up to eight times median incomes (Source, CML now UK Finance).

3.2.2 Government and other interventions in the market

As the affordability crisis deepened, and with the huge significance of the property market to the health of the wider economy, the government looked to interfere in the market by encouraging first-time buyers and dampening demand for buy-to-let and second home properties. They did this through punitive taxation changes, such as adjustments to Stamp Duty Land Tax (SDLT) and by incentives to encourage ownership, including shared ownership and help-to-buy schemes. These initiatives encouraged buyers back into the market, creating more activity in the building and professional services sector and releasing money back into the economy that would otherwise have been spent on rental payments and on saving for high deposits.

The **Help to Buy** initiative introduced in 2013 came in various forms such as an ISA, where saving is tax-free, and the government make a contribution to house purchase proportionate to the level of saving.

Shared ownership schemes, where part of the property is owned, and part rented.

Equity loan, being a scheme where the government has a 20 per cent share in the property.

These schemes are intended to reduce the deposit required, making house purchase easier and therefore stimulating the market.

Such schemes change, periodically, and for full details of current schemes, consult the government website.

As demand increased, the problems of an inelastic supply became more acute, particularly in London and the South East, where there is always a shortage of land for building. To address this, the government sought to encourage the release of brownfield sites and also to re-purpose redundant commercial units, by relaxing planning controls.

Moving forward, we started to see the impact of wider economic uncertainty on the housing market. In 2016, the UK held a referendum which resulted in an unexpected vote to leave the European Union (Brexit). After this, a sustained period of political uncertainty left homeowners reluctant to move and foreign investors reluctant to buy, until the economic impact of Brexit became clear. The market stagnated and dropped in some areas (particularly London and the South East).

As the markets began to recover from this shock, in 2019 an additional factor came into play, as the world faced a global COVID-19 pandemic. As the population were 'locked down' and offices were closed, many businesses and their staff were forced to work from their homes and ultimately to re-think their whole approach to working practices. With the restrictions of lockdown, and when the prospect of spending less time in the office in future became clear, the shock of a brief property market closure moved quickly on re-opening to a surge in demand as people began to rethink their requirements for their living space, looking for more room, outdoor space and a move away from the cities. As a result, the demand for rural and suburban properties surged and owners of city centre flats struggled to sell, causing prices to drop in some places. The impact of this was further amplified by another piece of government interference, aimed at stimulating the housing market and helping the economy post lockdown. The introduction of a temporary suspension to Stamp Duty Land Tax (SDLT) caused a surge in demand, as buyers sought to beat the 'cliff edge' deadline for this benefit imposed by the government. As we had seen before in 2016, this caused a flurry of activity, followed by a period of correction as the market re stabilised and found its true level.

3.2.3 Microeconomics and macroeconomics

As a closer to this section, it is worth pointing out that, while the macroeconomics of the wider economy and house prices are closely connected (along with the political interference that inevitably comes with it), the competent valuer must also have a good grasp of the local microeconomics influencing properties in the area in which they value. The impact on property values in any given area from factors such as factory closures, population movement, natural disasters and potential large infrastructure projects, to name but a few, must also be considered in any valuation decision. It is for this reason that it is key for any valuer to know their market and resist the temptation to extend into areas where the local economic drivers impacting prices are unknown. Working outside one's area of expertise, be that property type or location, will expose the valuer to a negligence claim.

! Benchmark: Economic climate

The benchmark here is the **economic climate** both locally and nationally. An example would be the earmarking for closure of one of the steel plants, either in Scunthorpe or Port Talbot. Local properties were selling quite well up until the potential for closure was announced, then the bottom dropped out of the housing market until the situation was clarified and confidence returned.

---

4 Law for valuers

An overview

4.1 Terminology

Before embarking on this outline chapter of legal matters, a brief explanation of three key phrases may be useful. A professional office should have recourse to good legal dictionary which will cover much more.

Reference: Dictionaries

A cheap paperback dictionary such as *Osborn's Concise Law Dictionary* (Woodley, 2013) is recommended for any professional bookshelf. Many online dictionaries are American, rather than UK, and can be misleading.

Civil law: broadly, this refers to the law of disputes between individuals and/or businesses (as opposed to the state). If the claimant does not bring an action, no-one else will. This might include contract claims, such as non-payment of debt, faulty goods or poor (negligent) service. It includes all areas of tort law, such as negligence, private nuisance and trespass. Civil law also covers family matters such as divorce and adoption; and the law relating to land and buildings – land law and tenancy.

Criminal law: although the victim, suffering personal injury or loss of money, may be an individual, criminal law can broadly be seen as wrongs where the state will (normally) seek to get involved. Although individuals *can* bring private prosecutions, it is more usual to bring the wrong to the attention of the relevant authorities for the matter to be taken further by the Crown Prosecution Service. The relevant authority will be a body of state such as the police, the Health and Safety Executive or the Environment Agency.

Common law: this term is sometimes used in contrast to legislation, i.e. where something is covered by case law only, and not an Act of Parliament. It is also used in a wider sense, in that the UK might be described as a common law jurisdiction in that we give court cases binding legal status on *future* cases, i.e. not all our law is covered by legislation.

DOI: 10.1201/9780367816988-4

4.2 Finding the law

For most non-lawyer professionals, the law is found in bulletins from professional bodies, such as the RICS or the CIOB, from the internet, from newspapers and other media, and from textbooks. But all these are just summaries and 'translations' of the primary sources of law which in the United Kingdom (and for these purposes we can include Scotland) essentially comprise legislation and case law. To access the primary sources when more detail is required or when a point needs to be clarified and to be able to support any statements or arguments you might make in practice, an understanding of how these sources work is vital for the competent professional. In addition, there may be opportunities to impact changes in the law. Hopefully this is not due to involvement in litigation as a defendant, but possibly through feeding in to calls for comment from Parliament, the Law Commission or a professional body, or as an expert witness.

For further detail on the law as applicable to the land and property sector, see *Galbraith's Construction and Land Management Law for Students* (7th ed., de Silva and Charlson, 2020).

4.2.1 Legislation

Legislation may also be referred to as statutes or Acts of Parliament. These are the most important sources of law in that they take precedence over court cases. The courts, even the Supreme Court, cannot overturn legislation. The courts can only interpret and apply. Whilst judicial interpretation can, in some cases, give a certain creative scope, the judges cannot contradict clear statements of law.

Judges can interpret where there is obvious scope in the wording of the legislation for the court to apply its view, such as in what constitutes 'reasonable care' in the Occupiers' Liability Act 1957 or 'reasonably practicable' in the Health and Safety at Work etc. Act 1974.

Judges can also interpret using common-sense and a view as to what the legislation was trying to achieve, or where there is a lack of clarity or ambiguity.

R v Allen (1872). Section 57 of the Offences Against the Person Act 1861 stated that 'whosoever being married shall marry any other person during the lifetime of the former husband or wife is guilty of an offence'. Interpreted literally there is a circular effect. Clearly as the second marriage is unrecognised at law, you cannot 'marry' twice. It was held that the statute should be read such that the word 'marry' should be interpreted as 'to go through a marriage ceremony'.

But where words are clear, then the courts would normally apply them as written, perhaps indicating that a change should be considered by Parliament.

In *Harrogate Borough Council v Simpson* (1984), the Court of Appeal had to consider the words 'spouse' and 'family' in the treatment of a same-sex partner seeking succession to tenanted property under the Rent Act 1977, as amended by the Housing Act 1988. The words were held to have clear meanings which did not include same-sex couples. This was overturned in part by *Fitzpatrick v Sterling Housing Association Ltd* (1999) where it was held that 'spouse' had a clear meaning and any extension was for Parliament, but that 'family' had a very different meaning in 1999 to that envisaged by Parliament even in the 1970s and 1980s, let alone when the original legislation referring to tenancy succession was passed (being the Increase of Rent and Mortgage Interest (War Restrictions) Act 1915).

4.2.2 Legislative process

Moving from an idea to an instrument of law which must be adhered to can be a long process. An Act of Parliament may have come from a variety of sources – perhaps something raised in the government manifesto, from lobby groups (including professional bodies), from individual Members of Parliament (maybe in connection with their own personal or professional interests or something raised by a constituent) or from Law Commission recommendations. Most legislation starts in the House of Commons but can start in the House of Lords. It will be formulated into a bill which will go through a process of scrutiny and, doubtless, editing in both houses. Only a minority of bills make it through this process but if they do, they need Royal Assent (albeit a formality). They are not operative law, however, until a date either mentioned in the act or, more often, as set by a commencement order. Some acts take many months or years to be fully implemented.

The Easter Act 1928, seeking to set the date of Easter Day, received Royal Assent in August 1928 but is not yet operative. No commencement date has been given due to continuing disagreement about the content of the act between different Christian denominations.

4.2.3 Delegated legislation

In addition to Acts of Parliament, which run to an average of 30 to 40 per year, there are other forms of law, referred to collectively as delegated legislation. The key forms of delegated (sometimes termed 'subordinate' or 'secondary') legislation are:

- **Byelaws**: provisions made by local authorities which can be viewed via the relevant local authority.
- **Statutory instruments**: provisions normally made by government departments. There are many thousands of these per year.

Reference: Accessing legislation

All Acts of Parliament since 1988, and key acts since 1267, along with statutory instruments, etc. can be seen at www.legislation.gov.uk.

4.2.4 Cases

As noted above, cases can only interpret legislation, not overrule it, so how can they be referred to as a *source* of law?

(a) that interpretation can have a considerable impact in the accepted meaning of a statute, and
(b) not all areas of law are covered by legislation.

There are considerable and important areas of law in this country where there are no operative Acts of Parliament and the law has been developed, often over centuries, into the current position. Examples include murder (you may have heard of the Homicide Act 1957 but that deals with defences to murder, not the offence itself). Closer (one hopes!) to professional areas of interest is the law of negligence, probably the subject of the greatest number and highest financial value of claims through the civil courts, all based on case law.

We must, however, be careful with the word case *law*. Judicial precedent operates, in simple terms, by judgments in the higher courts binding lower courts and usually courts at the same level, on future cases involving the same point of law. But the Supreme Court is never bound, so ultimately any case can be overruled, *and* there is often scope for 'distinguishing' a case, i.e. for it to be established that there are *material* differences between an apparently binding case and the case in hand. Cases are all decided on the facts of the case in hand so care must be taken not to quote as establishing incontrovertible rules applicable to all cases in future, as though they were a form of legislation.

4.3 Legal disputes

4.3.1 Courts and tribunals

When disputes cannot be resolved amicably or through preferable means such as Alternative Dispute Resolution, or where a crime has been alleged, then there will be recourse to a formal place of adjudication. This forum will vary depending on the subject matter in hand, and whether it is being heard for the first time or on appeal.

Civil courts and tribunals aim to right the wrong and, subject to exception, not to punish. That loss may be easy to quantify, e.g. a debt, or more difficult, such as personal injury, psychological damage or projected loss of earnings. Where financial compensation will not solve the problem, for example, where a persistent trespasser crosses land, then other remedies are available, such as an injunction – an order to stop (stop trespassing, stop building without permission, stop felling certain trees), or specific performance (an order to carry out a contract where the goods in question cannot easily be obtained elsewhere, so compensation would be inadequate). Remedies *other than* compensation, are available only at the court's discretion.

Criminal courts tend to concentrate on the punishment of the wrong-doer, albeit drawing in ideas of deterrence, rehabilitation and victim and societal retribution. They may award a fine, a custodial sentence or community service, and have other powers such as prohibition (e.g.. from running a business, attending an event or area or from keeping animals) or ordering drug rehabilitation.

There follows a summary of the main courts and tribunals likely to be encountered:

Supreme Court   Hears cases on all subject matter, including those on appeal from Scotland and Northern Ireland. Judgments must be followed by all lower courts, but the Supreme Court does not bind itself, i.e. it can reverse previous decisions (see *British Railways Board v Herrington* (1972)).

Court of Appeal   There are two divisions: civil and criminal. Judgments bind all lower courts and, normally, itself, subject to the provisions of *Young v Bristol Aeroplane Co. Ltd* (1944).

High Court   Largely a civil court (with a review function of the criminal courts). The court is split into three divisions (Queen's Bench, Chancery and Family) which have further divisions and courts. Note the Technology and Construction Court, the Commercial Court and the relatively new Planning Court, all within the Queen's Bench Division. All judgments bind the County Court; the court binds itself if a judgment is from more than one judge (i.e. a divisional court) but not if from a single judge. Judicial review is the process whereby the decisions of public bodies can be examined, e.g. the actions of government ministers, local authorities, the police, prisons, non-departmental public bodies, such as the Environmental Agency, and courts and tribunals.

Crown Court   Hears more serious criminal cases. Judgments do not bind other courts, but reasoned decisions and sentencing remarks are drawn up and may be persuasive in future.

Magistrates Court   Hears the vast majority of criminal cases and some civil matters such as non-payment of Council Tax. Most magistrates have no formal legal training, although are supported in court by a qualified lawyer. Some larger centres have qualified, full-time, paid judges in the Magistrates' Court. No judicial precedent is set at this level.

Family Court   This court was created by the Crime and Courts Act 2013 combining the family functions of the Magistrates Court and the County Court. The Family Court does not make judgments binding on future cases.

County Court   A civil court. The judgments are not binding on future cases but, as with the Crown Court, decisions are supported with written judgments which may be persuasive.

Tribunals   Tribunals have operated in the UK since 1911 but in their modern form since 1959, with the idea of taking certain civil matters out of court and into a specialist arena. Avoiding some of the strict protocols of court should help ensure that claimants can act for themselves, although this can be daunting, especially if the defendant is legally represented. Subject specialism means that there are people with professional backgrounds other than law. Typically, there may be a lawyer chairperson with subject specialist wing members, e.g. a trades union professional for an Employment Tribunal, a chartered surveyor for the Property Chamber. There are many different tribunals but most likely to be encountered are the Employment Tribunal, the First-tier Tribunal (Property) and First-tier Tribunal (Tax). All have appeals to, respectively, the

Employment Appeals Tribunal, Upper Tribunal (Lands Chamber, Tax and Chancery Chamber).

Tribunals do not make binding decisions (for future cases) either at first instance or on appeal but do produce reasoned judgments which are persuasive and often used in industry as a practice guide.

4.3.2 Alternative Dispute Resolution (ADR)

A professional person might well go through their career without being involved in a court case, and it is to be hoped that they do. It is almost inconceivable, however, that they would not encounter disputes which are resolved by alternative methods. 'Alternative' Dispute Resolution means alternative to court and embraces a number of forms of varying formality.

Arbitration Arbitration might be carried out either because terms of engagement or a contract indicate that disputes will be resolved by arbitration, or because the subject matter of the dispute has statutory reference to arbitration. This will involve the appointment of an arbitrator (who may have been agreed at the time of the contract, although not necessarily). The arbitrator may be a subject specialist, for example, have undergone the RICS arbitration qualification, or may be more general, possibly a member of the Chartered Institute of Arbitrators. The arbitrator is governed by the Arbitration Act 1996 and their decision is binding.

Expert adjudication This is similar to arbitration and is common in property disputes. It involves an appointed adjudicator who will come to a binding decision (subject to further recourse to formal arbitration or the courts). It is intended to be speedy (normally within 28 days), cost effective and with the aim of maintaining working relationships. Professional bodies often have their own protocols and frameworks, e.g. the RICS.

Mediation Mediation does *not* result in a binding decision, unlike arbitration. It involves a mediator, normally with specialist training, acting as a skilled facilitator between the parties. As well as a business or consumer context, mediation is also a key element of modern marital and civil partnership breakdown management, prior to divorce/dissolution.

Conciliation Conciliation is a word well known to those who lived through the 1970s and the years of industrial unrest with ACAS (originally the Advisory Conciliation and Arbitration Service) regularly in the news. Conciliation is not dissimilar to mediation, but may involve more guidance from the third party.

Reference: Employment law and advice

The invaluable source of employment information, advice and dispute resolution services, ACAS is helpful on all employment matters for both employers and employees and is particularly useful regarding dispute resolution: www.acas. org.uk.

4.3.3 Particular situations

Construction: Construction contracts can be particularly complex, and disputes are normally dealt with through a range of alternative procedures, largely agreed in the initial contract, which may include mediation, neutral evaluation, expert adjudication and arbitration. There can be formalisation of the outcome of alternative dispute resolution through the court settlement process in the Technology and Construction Court (part of the High Court). There are also dispute resolution procedures applied by default through legislation (including the Housing Grants, Construction and Regeneration Act 1996 and the Local Democracy, Economic Development and Construction Act 2009).

Party walls: Where a property boundary has a wall (whether in the garden, yard or as part of a semi-detached or terraced house or block of flats), then there may be need to encroach over the boundary to carry out works, or at the least, to cause noise or vibration. Or there may be a wish to make substantive alterations. Clearly the parties could liaise together to help ensure goodwill and no escalation of dispute. This is what normally happens, but there is a legal mechanism for agreeing ahead the works and dispute resolution procedures, through the Party Walls, etc. Act 1996.

Boundaries: Disputes involving the precise position of a boundary are not uncommon and often involve surveyors. Traditionally, if the parties cannot agree at an early stage, possibly taking on the advice of surveyors and/or solicitors, then the matter will go to court. This will be the County Court or High Court being a civil matter, or possibly the Land Registration division of the Property Chamber (First-tier Tribunal) if the matter involves Land Registry plans. This is almost never the best way to resolve things, given the undue costs.

See *Rashid and Akhtar v Sharif and Sharif* (2014) involving a dispute over 225 mm of land. The land was, anyway, inaccessible to the complainant party being behind a shed and the dispute ended in the Court of Appeal and the winning party *not* recovering their costs due to their behaviour being as unreasonable as that of the defendant.

Parties are strongly advised to consider arbitration or expert adjudication. Professional advisers should endeavour, from the earliest stages, to help clients resolve disputes without recourse to court. A bill was introduced into Parliament in July 2017, and again in January 2020, which gives the decision of appropriately appointed surveyors binding status, subject to appeal (Property Boundaries (Dispute Resolution) Bill (HL) 2017-19 and (HL) 2019).

Reference: Boundaries

Anyone working on boundaries should review the Land Registry Practice Guide 40 (regarding Land Registry Plans), sections 3 (*Land Registry: Plans*) and 4

(*Boundary Agreements and Determined Boundaries*) and the RICS *Practice Guide Boundaries: Procedures for Boundary Identification, Demarcation and Dispute Resolution* (3rd edn, RICS, 2014a) for valuable detail on both the law and practice of resolving boundary disputes.

4.4 Professional roles

Expert adjudicator: The use of an expert adjudicator is often found in construction contracts, but they are used in any technical area of dispute. They are experts in the field in question and reach a binding decision (unlike mediators, for example). There are no formal appeal procedures, but the dispute could be taken on to arbitration or court. The management of the expert adjudication process may be dictated by the standard form contracts as mentioned under Construction, above.

Expert witness: Where a dispute involves a point of technicality, as opposed to a point of law, then expert witnesses may be called. Traditionally both parties to a dispute would call their own expert and this is still common practice. It is vital, however, for the expert in question to be aware that their duty is to the court and *not* to the party paying their fees. This means that they do not have a role in supporting a particular side to the case, but rather that they simply state without bias their view of what they are being called on to explain.

It is sadly common to see expert witnesses in court being chastised for being too partisan, not reporting in the required format and being unprepared (see, for example *Van Oord v Allseas UK* (2015) where Coulson J in the Technology and Construction Court found no fewer than 12 express reasons for finding an expert witness's evidence 'entirely worthless'.

Hubbard v Bank of Scotland (2014) was a negligent mortgage valuation case in the Court of Appeal. It was asserted that cracks in a wall should have alerted the valuer to possible subsidence, both reducing value and indicating an express direction to the purchaser to have a full structural survey. The expert witnesses were asked, on the basis of contemporary photographs, the original valuer's report and a current inspection of the property, to say whether the valuer's conclusion (that the cracks were old and not evidence of imminent subsidence) was reasonable. The judge preferred the expert who concluded that the original valuer *had* made a reasonable decision. Importantly, the other expert witness did not have weight put on their conclusions not least because of how they conducted their work, for example, making it clear that they had not read all the information.

The Society of Expert Witnesses and professional bodies hold registers and run training courses, but there is no express qualification requirement. It is generally a matter of individuals putting themselves forward as being available for this work or, where someone has built up a particular area of expertise and is known in that context, being approached to act as an expert. There are, however, ethical requirements from some professional bodies. The RICS Practice Statement, *Surveyors Acting as Expert Witnesses* (4th edn, RICS, 2014b) stipulates that surveyors must not accept instructions to act as experts in areas outside their expertise (3.10).

Reference: Expert witness work

If undertaking expert witness work, then the following documents are required reading.

Civil Procedure Rules (Practice Direction 35 – Experts and Assessors) or Criminal Procedure Rules (Part 19 – Expert Evidence), as appropriate.
Professional guidance from a relevant body, e.g. RICS, *Surveyors Acting as Expert Witnesses* (2014b), see also *Surveyors Acting as Expert Witnesses, Client Guide* (RICS, 2014c).

Solicitors and barristers: Solicitors and barristers are, of course, legal professionals in England, Wales and Northern Ireland with the respective terms solicitors and advocates being used in Scotland. This distinct separation of classes of lawyer is relatively unusual in the world at large and now only really exists in Hong Kong, South Africa and parts of Australia. Lawyer, by the way, is a general term for someone who works in the law. It does not denote any particular qualification although would tend to signify either a solicitor or barrister in this country.

Most countries only have one form of qualification for legal professionals. Traditionally there was a greater separation of work than there is now, in that barristers were thought of as the court lawyers with solicitors in offices dealing with clients. It is still the case that the general public and most professionals will first call on a solicitor for legal advice. Professionals should know that there is the possibility of going direct to a barrister (see the Direct Access Portal of The Bar Council website), but this is not particularly common.

Solicitors are sometimes seen as being rather like a GP, in medical terms, who will then call on a specialist (barrister) if required, either for their additional expertise or court appearance. But even for what are termed High Street solicitors dealing with the general public, few nowadays are 'GPs' in that they tend to specialise. The same solicitor is not likely to be dealing with conveyancing, divorce, probate, business tenancies, employment and whatever other issues may arise. Firms, or departments and individuals within firms, will specialise. Professionals should certainly ensure that they seek legal advice from specialists and ensure they are the precise specialist they need, for example, if you have a problem with a residential tenancy, the best advice will not be obtained from a commercial tenancy specialist.

Barristers still do the bulk of court work (although some barristers do not, and work in-house for companies or organisations as legal advisers) but solicitors can work in the lower courts and can do an additional qualification to become solicitor advocates.

To qualify, both solicitors and barristers start with a qualifying law degree *or* any honours degree plus a one-year law conversion course. Then the paths diverge and there is an additional year with different courses for barristers and solicitors. After the aca-demic elements are passed, solicitors require a two-year training contract (formerly known as articles), whereas a barrister requires one-year pupillage. This is a structured period of full-time work under the supervision of a qualified solicitor or barrister, which leads up to full-qualification and gaining the practice certificate required to undertake the full respective roles of solicitor or barrister and which must be renewed annually. Solicitors apply to the Solicitors Regulation Authority. Barristers apply to the Bar Council through the Authorisation to Practice.

5 Law for valuers: land, contract and tort

Land is, of course, central to the residential property valuer's role. And here is a review of land tenure, third party interests in land, and other pertinent issues. These issues can be of direct and vital relevance to valuation – how can one value without knowing if land is held freehold or, if leasehold, how long is left on the lease? How can one value land for development, without understanding the significance of a restrictive covenant in place?

5.1 Land

Property law relates to ownership and other rights regarding 'things'. It is important, as a fundamental, to define the word 'land' as the law treats land, fixtures (part of the land) and chattels (not 'land') differently. You also need to know what the limits of land ownership are, what degree of control you can exercise over your land and to what extent you must recognise the rights of others. A definition of 'land' helps to clarify all of these issues.

A legal definition of land is set out in the Law of Property Act 1925 interpretation section, s205 (1) (ix). This states:

> Land includes **land of any tenure**, and **mines and minerals, whether or not held apart from the surface, buildings or parts of buildings** (whether the division is horizontal or vertical or made in any other way) and other corporeal **hereditaments**; also a **manor**, an **advowson**, and a **rent** and other **incorporeal hereditaments**, and **any easement, right, privilege or benefit in, over, under or derived from land**; but **not an undivided share in land** …

The following notes add some explanation to the phrases above emboldened by the author.

5.1.1 'land of any tenure'

'land of any tenure'

There are only two legal estates in land:

Estate in fee simple	freehold	realty
Term of years absolute	leasehold	personalty

DOI: 10.1201/9780367816988-5

Land may be held (have the tenure of) either **freehold** or **leasehold**. People most often associate the word tenure with land that is held leasehold, but land held freehold is also held under a form of tenure. To explain this, one has to travel back to the arrival of the Normans in England, from the middle of the eleventh century. At this time, all land was considered to belong to the Crown. The King granted land (often referred to as manors) to his lords who, in turn, would grant land to others (a process known as sub-infeudination). Land that was granted **freehold** gave the tenant the right to use that land in perpetuity. In return for this grant, the tenant owed his landlord, ultimately the King, certain services. These were knight service (a duty to provide a certain days per year armed service and/or a number of armed horsemen), frankalmoign (religious service of praying for the soul of the King or lord), and socage (the duty to provide a certain number of days work on the land).

Over time these duties were commuted to monetary payments. Time then eroded the value of these payments until they were no longer worth collecting and disappeared and so we tend to forget that land held freehold is technically tenanted from the Crown or another superior landlord, but that is still the position. No one actually *owns* land (except the Crown). What is actually owned is a bundle of rights to use the land. This bundle of rights is known as an **estate**. Prior to 1925 many different types of estate existed but the Law of Property Act 1925 destroyed all but two, leaving us with the modern freehold and leasehold.

Freehold

The **freehold estate** is the most valuable estate that can be owned and is known, legally, as a 'fee simple absolute in possession'.

fee   An inheritable estate.
simple   Inheritable by anyone.
absolute   An unconditional, unmodified estate that will go on forever rather than being limited by an event which may occur in the future, e.g. a marriage.
in possession   The owner is either in physical possession or is entitled to rents from the land.

Essentially the fee simple is a perpetual right to land.
Note the existence of the following:

fee tail   Limiting the succession of property. No new entails may be created since the Trusts of Land and Appointment of Trustees Act 1996.
life estates   Passing an estate in land for the length of beneficiary's life, or for the length of another's life, i.e. *pur autre vie.*
commonhold   Under the Commonhold and Leasehold Reform Act 2002, which now allows flats to be held in freehold with the communal areas, such as halls and stairs, to be held by the commonhold association.

Note, also, statutory overriding of the extent of vertical ownership, such as rights relating to minerals noted below. The common law concept, as restated by the seventeenth-century jurist, Edward Coke (1628), was that: 'He who owned the land surface owned everything up to the heavens and down to the depths of the earth.'

Over time, statutes and the courts have placed certain practical limits on this with the land holder retaining exclusive control of the immediate atmosphere.

Cases of trespass, or otherwise

Kelsen v Imperial Tobacco Co. (1957) indicated that projecting advertising hoarding = trespass.

Woollerton and Wilson v Richard Costain (1970) found a projecting crane jib = trespass.

Bulli Coal Mining Co. v Osborne (1899) and *Star Energy v Bocardo SA* (2010) held that tunnelling under a neighbour's land = trespass.

Bernstein v Skyviews & General Ltd (1977) found that aircraft passing over at a 'reasonable' height' ≠ trespass.

Leasehold

For a leasehold to exist, the law requires:

1. certainty of duration (section 1(1)(a)) and 205(l)(xxvii) Law of Property Act 1925)
2. exclusive possession
3. payment of rent

Street v Mountford (1985) set out these requirements clearly, along with a review of the supporting preceding law dating to the thirteenth century and should be read by anyone wanting to gain a sound grounding in the nature of leases.

Certainty of duration

The **leasehold estate** is an estate which is held for a 'term of years absolute', i.e. a specified period which must be set out at the start of the tenancy (although this can be subject to extension on a rolling periodic basis). The tenant owns the right to use the land for this period in return for paying rent to the landlord.

The period must be certain, e.g. a lease cannot be granted for a person's life because the duration will be unascertainable. Other uncertain periods found to fall short of leasehold requirements include:

- 'until the land is required for road widening', *Prudential Assurance v London Residuary Body* (1992),
- 'as long as the company is trading', *Birrell v Carey* (1989), and
- 'for the duration of the war', *Lace v Chandler* (1944).

Contrast freehold, where the period is either indefinite (fee simple) or uncertain (fee tail/life estate).

Exclusive possession

If there is no exclusive possession, there is likely to be a **licence** which does not afford the statutory protection for tenants.

Retaining an express right of entry will be fatal to finding a lease, e.g. where a city corporation let a dock but retained the right to enter the land to open and close dock gates and to clean it, this could not, due to the freeholder's right to enter, be a lease (*Wells v Kingston-upon-Hull Corporation* (1875)) . Lodgers will normally have a licence rather than a lease (*A G Securities v Vaughan* (1988)). However, the landlord simply retaining a key does not negate the existence of a lease (*Family Housing Association v James* (1990)). Importantly, the courts will look at the facts and the content of the agreement rather than the parties' intentions (*Street v Mountford* (1985)).

Note: different land usage is governed by different statutes with regard to leases – the following are major statutes which will be encountered but there are others:

Residential   Rent Act 1977, Housing Acts 1988 and 1996, Leasehold Reform Act 1967, Commonhold and Leasehold Reform Act 2002
Commercial   Landlord and Tenant Act 1954, Part II
Agricultural   Agricultural Holdings Act 1986. Agricultural Tenancies Act 1995

5.1.2 Other interests in land

'mines and minerals, whether or not held apart from the surface'

This second part of the definition of land is fairly self-explanatory and tells us that it is possible to sell and hold mineral rights separately from the land above them. In fact, the state owns all coal (Coal Industry Act 1994), oil and natural gas (Petroleum Act 1998) with the Crown owning gold and silver (per *The Case of Mines* or *R v Earl of Northumberland* (1568)).

'buildings or parts of buildings (whether the division is horizontal or vertical or made in any other way)

This part of the definition confirms that buildings are considered to be land and that whether a flat is on the ground floor or a penthouse, it will still be treated as land in the same way. Note that land can be split horizontally by means of a flying freehold. This is often seen in older properties where upstairs and downstairs walls do not line up or where a room overhangs a drive. This will require rights of support and access and needs to be handled properly during the conveyancing process.

This also leads on to the point that items that are attached to buildings or land will become part of that land. These items are known as fixtures.

A fixture must be distinguished from a chattel, which can be freely removed from a holding, thus the matter is of key importance on sale.

Although most questions should be solved by modern conveyancing practice and the forms vendors are required to complete, specifying what will be left or taken from a property on transfer, there is still scope for disagreements regularly reaching the courts. When deciding whether an item is a fixture or a chattel, the courts will consider the degree of annexation (whether it is physically attached to the land) and purpose of annexation (why it has been brought onto the land), per *Holland v Hodgson* (1872).

A conveyance of land passes land and fixtures (Law of Property Act, section 62(1)) unless specifically excluded in the contract.

A contract for the sale of land must, bar very few rare instances, be in writing per Section 2 of the Law of Property (Miscellaneous Provisions) Act 1989. This requirement of writing does not affect the operation of a resulting, implied or constructive trust, a complex area of law based on promises and/or intentions, which is beyond the scope of this book.

'corporeal hereditaments'

'Hereditament' means land which is inheritable with 'corporeal' indicating that it has physical presence, e.g. land and attachments thereon, such as buildings, fences, trees, statues, turf, etc. Thus, essentially, tangible real property.

'manor'

Manorial rights are largely a relic of feudal times, but some still exist. They may include the right to call oneself Lord of the Manor and, rather more importantly, mineral rights, the right to license fairs and markets and possibly duties, such as to clean ditches.

'advowson'

The right to nominate the next Church of England vicar or rector, often laying with the manorial rights. Technically 'land' but not land capable of registration.

'rent'

In this context, a periodic payment, secured on land, e.g. all properties in a development paying an annual charge to an estate company for maintenance, or a child is left land in a will by the father subject to paying £1,000 per annum to the mother. They may be payable by a freeholder and are to be distinguished from the ordinary concept of rent payable by leaseholders or licensees. Rentcharges, as they are termed, are largely localised, particularly in the Bristol and Manchester areas. Creation of new rentcharges was abolished with the Rentcharges Act 1977, with any still in existence to be extinguished in 2037.

'incorporeal hereditaments'

Largely, intangible real property which is treated as part of land, e.g. rentcharges, easements and profits (see below).

'an easement, right, privilege or benefit in, over, under or derived from land'

This part of the definition of land deals with third party *interests* in land as opposed to first party freehold and leasehold *estates* in land. Such interests can be legal or equitable:

Legal interests in land

- Certain easements and *profits à prendre*
- Rent charges
- Certain mortgages
- Certain rights of entry annexed to a lease
- Tithes and charges (abolished)

Equitable interests in land

- Trusts
- Mortgages
- Restrictive covenants
- Entails
- Life estates

It should be noted that while equitable interests are recognised, they are not as valuable as legal interests. A legal interest is good against the whole world whereas an equitable interest is good against everyone except 'equity's darling': the *bona fide purchaser for value without notice*. This is a person who buys in good faith not knowing about an interest and who could not have found out about it by making routine inquiries. If such a person buys property, then third party equitable interests are lost. In practice, the problem of the *bona fide purchaser* is overcome in two ways:

1. By registering equitable interests. Once these interests are registered, any prospective purchaser is considered to have notice that they exist (whether or not they or their professional advisers have actually seen the register).
2. By making beneficial interests under trusts *overreachable* (this means that money is paid to the trustees to hold for the beneficiaries so that the purchaser can take the land free of the equitable interest). Only a very few equitable interests remain subject to the *doctrine of notice*, and the developments in land registration since 2002 (Land Registration Act 2002) mean that it will be of decreasing relevance as time goes on, but such interests do include restrictive covenants created before 1926.

Legal third party interests which are recognised by the Law of Property Act 1925 are (for practical purposes):

Easements: A right over one piece of land, which benefits another piece of land and is attached to that piece of land, e.g. a private right of way, of which more below.

Profits à prendre: A right to take natural resources from another's land, such as grass, minerals, timber or wild game. Profits are similar to easements but with a right over land which diminishes it in some way. This right may be attached to another piece of land (known as *appurtenant*). Such rights *not* attached to beneficial land are termed *profits à prendre in gross*.

Rentcharges: Periodic payment of money, secured on land, which is not a rent or a mortgage (see above).

Charges by way of legal mortgage: Transfer of a legal interest in land for the purpose of securing the repayment of a debt.

All these legal interests are good against any purchaser of land, i.e. the new purchaser must recognise them even if they do not come to light as a result of inquiries made in the process of a sale.

NB: These legal third party proprietary interests in land, which are binding on those who acquire the freehold or leasehold of the land, must be distinguished from personal interests, such as licences, which are not binding on successors in title and are simply agreements between the parties to the original transaction.

5.1.3 Easements

Easements may be briefly defined as:

- irrevocable
- private rights
- enjoyed by successive landowners
- in respect of neighbouring land.

This highlights the importance of the classification, i.e. if the right is an easement, it passes with the land on transfer. If a right is not an easement, it may be a licence, which would simply be a matter between two parties and would not bind successors.

To be classified as an easement, it needs to enjoy certain characteristics. If these are missing, they are not capable of being an easement.

Once it has been established, due to the presence of these characteristics, that the right is *capable* of being an easement, it then needs to be established whether or not the easement has been created.

The requisite characteristics were clarified in the case of *Re. Ellenborough Park* (1956), which involved the communal use of parkland attached to a row of residential properties in Weston-Super-Mare:

- There must be dominant (benefitting) and servient (burdened) land.
- The easement must benefit (accommodate) the dominant land, i.e. be directly related to the enjoyment and normal use of the dominant land. Land need not be adjoining although any great distances are unlikely.
- The dominant and servient land must be separately occupied. The two plots may, however, have one owner with one or both plots having tenants.
- The right must be capable of forming the subject matter of a grant. The easement must be sufficiently definite. If it is too wide or vague, it will fall foul of this provision, thus there can be no general right of drainage, only a right through a defined channel (*Palmer v Bowman* (2000)). Similarly, although there is a right to light (subject to the technicalities of this area of law), there can be no right of easement to a pleasant view (*Aldred's case* (1610)). The courts are very wary of creating new classes of easement if they restrict the servient owner's enjoyment of land.

It has long been considered that there must be no exclusive or joint enjoyment of the land (*Batchelor v Marlow* (2001)) , although cases involving the parking of vehicles have questioned this principle (see *Moncrieff v Jamieson* (2007) and *Kettel v Bloomfold* (2012)). But the easement must be compatible with the servient owner's possession.

There must be no expense for the servient owner. In addition to any on-going or periodic expense being detrimental to the existence of an easement, the servient owner must not be bound to carry out repairs in allowing the dominant owner to enjoy the easement (subject to exceptions in an agricultural context, per *Crow v Wood* (1970)).

So once it has been decided that the right being sought is *capable* of being an easement, the question of creation must be addressed.

Creation of easements

Express: This is the most straightforward means of creation and there will be a clear indication in a deed or Land Registry documentation. This may be made at any time or on a sale of land by grant (giving land sold rights over retained servient land) or reservation (retaining rights over land sold for the enjoyment of the retained dominant land).

Implied: An easement can be implied where:
- there is an easement of **necessity**, e.g. to access retained land,
- evidence of **common intention**, e.g. where both buyer and seller of land know the usage to which land will be put, or
- where **the rule in *Wheeldon v Burrows* (1879)** operates. This is where a part of a land holding is sold (being the dominant tenement), and the grant of certain easements is implied where they are: continuous and apparent (generally taken to mean rights which are operative all the time, such as a right of light, rather than rights only utilised periodically, such as a right of way); necessary to the reasonable enjoyment of the property (wider than absolute necessity); have been used by the owner of the entirety for the benefit of the part granted, i.e. the dominant tenement is divested.
- where **Section 62 Law of Property Act 1925** operates, such that an easement can be created on transfer where existing rights are deemed to pass with the land.

Prescription

Where there has been continuous, long-running use, as of right, then an easement may be deemed to have been created by common law (usage since 'time immemorial'), by lost modern grant or under the Prescription Act 1832.

The **common law** presumes that an easement can be acquired by proof that it has been used to date since time immemorial (deemed to be 1189!). Clearly this is impractical, so evidence within living memory is accepted, but proof that it could *not* have been used since 1189 will defeat this.

The legal fiction of **lost modern grant** steps in to say that if the easement has been used for 20 years *at some point* since 1189, the law assumes that there was an express written grant/permission, but it has been lost.

The **Prescription Act 1832** tried to solve these problems such that 20 years usage, openly, without force and as of right (i.e. without permission), immediately before the claim, will suffice. And if there has been permission, then 40 years continuous and uninterrupted use is required.

The complications associated with this have been reviewed by the Law Commission (reporting in 2011) and it is hoped that legislation to simplify the creation of easements will be introduced in time.

Other than by agreement, it is not simple to get rid of an easement. It may be that a substantial change of use is sufficient (see *Attwood v Bovis Homes* (2001) and *McAdams Homes v Robinson* (2004)) although mere non-usage (in the absence of some expression of intention) is not enough (see *Benn v Hardinge* (1993) where 175 years of non-usage had not extinguished the easement).

Reference: Easements claimed by long usage

The Land Registry Practice Guide 52 – *Easements claimed by prescription* is useful reading.

5.1.4 Restrictive covenants

A restrictive covenant is a promise not to do something, relating to a certain piece of land. The promise must be contained in a deed or at Land Registry, i.e. it can only be expressly created (in contrast to easements, which can be implied), and must touch and concern related land, e.g. a promise not to build on one piece of land must benefit another piece of land close by. Restrictive covenants can be vaguer in terms than easements, e.g. a right not to have the view obstructed (*Wakeham v Wood* (1981)).

They can be modified or discharged under the provisions of section 84, Law of Property Act 1925 on the grounds that:

*   the character of the property or neighbourhood has altered;
*   the covenant impedes reasonable use of the land for public or private purposes;
*   the proposed discharge or modification will not 'injure' those entitled to the benefit of the covenant;
*   there has been express or implied agreement to the discharge or modification.

The principle that restrictive covenants run with the land (as opposed to agreements between parties) was established in *Tulk v Moxhay* (1848) which involved a covenant to keep Leicester Square, London, in 'an open state', which is still operative today.

A valid restrictive covenant will operate regardless of planning permission, so care should be taken to ensure the existence of such on land *prior* to expending monies on the planning process or valuing property with planning permission only to find developments impeded by the existence of a covenant. A retrospective claim can be made when building work has taken place in contravention of a covenant, but discharge is never certain and, at the least, may require the payment of substantial compensation.

'... but not an undivided share in land ...'

There are two ways in which land can be held by more than one person (both are referred to as forms of co-ownership). It should be noted that the terminology of tenancy has no relevance for the nature of legal estate held (freehold or leasehold). One is a joint tenant or tenant in common of freehold *or* leasehold land.

Joint tenancy

This is characterised by four unities:

Unity of Possession	Concurrent rights to enjoy the whole of the land that is owned.
Unity of Interest	Each co-owner must have the same estate and equal rights to the land or the proceeds from it.
Unity of Title	The title of each co-owner must be derived from the same document or transaction.
Unity of Time	Each joint tenancy must commence at the same moment e.g. one piece of land left to two daughters to inherit when they each reach 21 years old would *not* satisfy the unity of time.

Tenancy in common

Each tenant owns a separate yet undivided share of the land. The unity of possession must be present but at least one of the other unities is missing (often the unity of interest).

The two types of co-ownership also differ in another way: the right of survivorship and wayleaves.

The right of survivorship

- In a joint tenancy, property will pass on death to any surviving joint tenants. This is not overridden by a will, unless that will expressly severs the joint tenancy.
- In a tenancy in common, property will pass under the normal laws of succession, i.e. as stipulated in a will or under an intestacy.

It is not uncommon for parties to be unaware as to how they hold, or of the implications, either at the outset or, for example, where one party to a joint tenancy severs with the parties now holding as tenants in common (see *Boycott v Perrins Guy Williams* (2011)). Whilst competent lawyers should make the matter clear on property acquisition or when changes occur, it is prudent to ensure the right questions are asked. It is likely that, without any expression of preference on acquisition, co-habiting couples will hold as joint tenants and business property will be held as tenants in common.

5.1.5 Wayleaves

Wayleaves are temporary rights granted annually or for a given period, such as are acquired by utility companies, e.g. regarding pipelines or power cables. Note that some utilities companies may establish rights by way of easement, rather than the contractual wayleave mechanism.

5.1.6 Boundaries

Unregistered land will have descriptions and plans in the conveyance which will normally be conclusive but may also be unclear.

Registered land relies on Ordnance Survey maps filed against the title number in the Property Register. These may not be particularly clear, and it is important to note that Land Registry plans are, per section 60 Land Registration Act 2002, 'general' only. That is, they are not definitive. The 'determined boundary' procedure enables the exact line of a boundary to be shown on the register. Only if the procedure to determine a boundary has been met, or there has been a ruling after a boundary dispute, will the consequently lodged plan be a precise indicator.

In *Simpson v Sandford St Martin Parish Council* (2018) Cook J noted her surprise that qualified chartered surveyors on both sides of a dispute seemed unaware of the non-definitive status of Land Registry plans.

That said, they are, of course, a sound starting point and may well resolve disputes. In addition to these plans, there may be other evidence from older plans retained in deeds, and clues apparent from a physical inspection of the property involved.

If there is no clear evidence in the form of plans, then auction particulars, photographs, surveyor's evidence and acts of ownership (such as the erection and maintenance of a hedge, before any dispute) may all be employed.

There are a number of rebuttable presumptions used to try to establish boundaries:

Hedges and ditches   Where a hedge and ditch run together, the border is said to be on the edge of the ditch furthest from the hedge, subject to the ditch having been dug when the two sides were in separate ownership (*Vowles v Miller* (1810), *Wibberley v Insley* (1999)).

Hedges   Where there is a hedge alone, the border is said to run in the middle of the hedge. (Although ascertaining the precise middle of the hedge has given rise to more than one neighbour dispute!)

Roads   The mid-line of any boundary road is used (*ad medium filum*) although roads are often specifically retained by developers and for adopted highways, rights and obligations of access and maintenance vest in the local authority.

Rivers   Non-tidal waters – the boundary will be said to lie in the middle of the river (*ad medium filum*), giving adjoining landowners fishing and water abstraction rights

Tidal waters – the line is drawn at median highwater mark, the foreshore being held by the Crown.

5.1.7 Registered land

A system of registering the title to land was introduced in England and Wales in 1862 (having existed in Middlesex since 1709 and in Scotland since 1617). From 1990 there was compulsory registration triggered by certain events such as the transfer of a freehold or the grant of a lease of over 21 years. The Land Registration Act 2002 increased the aim of total registration by reducing the leases requiring registration to 7 years and having the requirement for certain third party rights to be registered, such as mortgages. The Act also introduced electronic conveyancing with the aim of speeding up property transfers.

The register is open to the public. You can now obtain historical copies of the register that Land Registry holds in electronic form. You can obtain a copy of any document submitted to Land Registry after 13 October 2003 unless it is especially exempt.

Disclosing unregistered interests

Be aware that not all interests affecting land are shown on the register. Now, when you register land for the first time or ownership changes, some unregistered interests must be disclosed to Land Registry. Examples include certain short leases and rights of way.

Unregistered land

Land not in the Land Register is 'unregistered' and title is found in deeds. The body of unregistered land is decreasing daily as the triggers for unregistered land noted above, such as transfers of freehold or longer leases, or mortgages, bring more property onto the Register.

5.1.8 Adverse possession (squatters' rights)

A squatter on registered land can apply to be registered as owner of that land if he has been in adverse possession (i.e. been squatting there unchallenged) for at least 10 years. The likelihood of the squatter's success has been diminished under the provisions of the Land Registration Act 2002 in that the paper owner will be notified of such an application, have the opportunity to object and, if the squatter does not vacate, can instigate County Court possession orders.

Where the land is unregistered, the old regime under the Law of Property Act 1925 and the Limitation Act 1980 applies and the squatter can apply to the Land Registry after 12 years and the original paper title holder will have no recourse (subject to the squatter being able to evidence that they have fulfilled all the requirements necessary).

It should be noted that, whereas the above is all a matter of civil law, squatting in residential premises is a crime, thus (a) the police can be called on to assist with removal and (b) it is subject to punishment by fine or imprisonment, per the Criminal Law Act 1977 and Legal Aid, Sentencing and Punishment of Offenders Act 2012.

Reference: Land law

With most areas of land law covered above, the Land Registry Practice Guides are an invaluable and very accessible source of reference for the practitioner: www. gov.uk/topic/land-registration/practice-guides.

5.2 Contract law

5.2.1 Basic structure

It is hardly necessary to say that contract law impacts every aspect of business and personal life. What may be a surprise to readers is that the basic structure of a contract is the same whether the subject matter is a child buying sweeties or a multi-million pound international trade deal. What differs are, of course, the terms and the detail in which those terms are set out. But before considering terms, the basic framework of a contract should be understood.

Offer

There needs to be an offer. This may be made in terms of an offer of goods or services for sale, or an offer to purchase those goods. For example, items in a shop window or being sold at auction are simply an invitation to treat, that is a pre-contractual expression of the potential for contract. So the potential purchaser makes an offer to purchase.

Acceptance

The offeree (the vendor in our example, the shop or the auctioneer) can accept or reject that offer. For example, they may know the offeror is under age when trying to purchase cigarettes, or the auctioneer may have two all but simultaneous bids and clearly cannot accept both. So an offer which has been validly accepted tends to indicate the point at which a contract is agreed and binding.

Consideration

The validity of the agreement is subject to there being good consideration. Consideration can be regarded as what is given in exchange, which, for the vast majority of contracts, will be goods or services in exchange for money, the goods or services being consideration for the money, the money being consideration for the goods or services. There is no legal requirement that the consideration is of equivalent value, thus a contract for £1 per year annual rent is valid (assuming no fraud is involved), but there must be *some* value. Where there is nothing of value given in exchange, the agreement will not, normally, be legally binding and thus not actionable in the event of a problem.

Intention to create legal relations

Even if a valid offer, properly accepted and supported by good consideration is found, there must be an overarching intention to create legal relations. In a commercial context this is almost impossible to deny, but in a social or domestic setting, it may be established that an agreement was simply that: social or domestic, with no intention that there be a binding contract with recourse to court in the event of a dispute. Family and friends can, of course, enter binding contracts and this will be apparent from the situation. If in doubt, ensure that written terms are set out, for example, using a lottery syndicate form (available from High Street stationers) to formalise lottery entries with groups of friends.

Even where the above elements are in order, the contract may not be valid due to a defect.

> **Form**: for most contracts, a verbal agreement is binding (subject to scope for disagreement over what was said) but certain contracts are legally required to be in writing. Transactions for the transfer of a freehold, leasehold, interest

or charge over land are to be in writing and signed by both parties (Law of Property (Miscellaneous Provisions) Act 1989, s2).

Capacity: if a party to a contract does not have full legal capacity to act, the contract may be unenforceable. In short, this will include minors (for certain matters), those lacking mental capacity and companies, although for companies there are few constraints to capacity since the Companies Act 2006 reforms. In practice, the only anecdotal evidence known of this area of contract law being an issue is with regard to dealing with minors at auctions who will not be bound under the contract, although any adult party to the contract will be. Minor can ratify contracts made before, on reaching the age of 18.

Legality: if the subject matter of the contract is unlawful, the contract is void.

Defects: if a party entered a contract through **misrepresentation** (i.e. being told an untrue statement of fact which was material to the decision to enter the contract), through **mistake** (a complex area which goes beyond individual lack of understanding), through **duress** (physical threats) or due to **undue influence** (either actual or presumed, based on relationships of unequal power), then the contract may be void.

5.2.2 Terms

Having established that there is a valid contract (i.e. an *enforceable* agreement), when things go wrong, what exactly can be argued? A court case or, ideally, less formal dispute resolution mechanisms, will be based on one party claiming that the other has breached the contract or, more precisely, a term (or terms) in that contract. These terms can be oral (subject to legislative provisions for certain contracts such as transfer of real property), written, included in a formal document headed 'Contract', in additional written representations such as related letters or emails which may be deemed to be part of the contact, or implied into a contract through legislation (for example, the Sale of Goods Act 1979 and the Consumer Rights Act 2015, with regard to contracts for the sale of goods, and the Employment Rights Act 1996 and other statutes, with relation to employment contracts). Terms can also be implied through course of dealing where terms were agreed in the past and the contract in question is one in a long-standing business relationship.

In a valuation context, it is important that terms are clearly set out and that they are understood by clients. For example, does the client understand the difference between a mortgage valuation and a survey? Do they understand what will and will not be assessed by the valuer? It was stressed in *Hart v Large* (2021) that simply giving the name of an RICS product was not enough – the details, and limitations, of the work being undertaken must be explained to the client.

5.2.3 Estate agency and surveying practices

The estate agency and surveying industry have a number of particular contractual considerations to bear in mind. These include provisions which operate when a contract is made in the client's home, as opposed to the business office; and also being aware that provisions will vary according to whether the client is a consumer or another business.

Reference: Estate and letting agency guidance

The National Trading Standards Estate and Letting Agency Team is run by Powys County Council (lead on the Estate Agency Act 1979) and Bristol City Council (lead on the Tenant Fees Act 2019). Their booklet *Guidance, on Property Sales* (updated 2019), is useful reading for estate agents: www.bristol.gov.uk/documents/3368713/0/NTSEAT+guidance+on+property+sales+July+2019.pdf/5c4d2d7b-b083-1f3c-309d-ba8caed5265c.

5.2.4 Off premises contracts

Where a contract is made in the office, then it will be binding on both parties from the date of the agreement. Where a contract is made with a consumer in their own home, then it is subject to the Consumer Contracts (Information, Cancellation and Additional Charges) Regulations 2013. This legislation includes a range of information which must be provided in such a contract and also increased the previously operative 7-day cooling-off period to 14 days, i.e. if a consumer changes their mind about entering the contract within 14 days, they have every right to cancel. This is obviously not helpful if clients have asked, for example, for a valuation to be completed without delay. The right to cancel can therefore be waived by the inclusion of a term in the contract indicating the waiver. As a significant term removing rights from the consumer, it is not sufficient that this is in small print buried in the middle of a range of standard terms. It must be expressly brought to the client's attention and would normally be seen as a separate item with a tick box or signature required.

Off premises contracts to include:

- **details of the services** being provided;
- the identity of the trader/the trader's **trading name**;
- the trader's **address**, telephone number, fax number and email address; the address and identity of any other trader you are acting on behalf of;
- the address of the business where the consumer should address any **complaints**, if different;
- the total **price** of services inclusive of taxes, or the manner in which the price is to be calculated;
- all **additional charges** and any other costs or the fact that such additional charges may be payable;
- the arrangements for **payment** and the **time** by which the trader undertakes to perform the services;

Be very clear on the figure/calculation and payment terms of commission.

- where applicable, the trader's **complaints** policy;
- time limits for **cancellation**, i.e. 14 days;

- that consumer will be liable if express request is made for the trader to carry out **work before the end of cooling-off period**;
- in the case of a sales contract, a reminder that the trader is under a legal duty to supply services that are in conformity with the contract, i.e. normal **contract law** applies;
- where applicable, the existence and the conditions of **after-sale** customer assistance, after-sales services and commercial guarantees;
- the existence of relevant codes of conduct, as defined in reg. 5(3)(b) of the **Consumer Protection from Unfair Trading Regulations 2008**, and how copies of them can be obtained, where applicable;
- where applicable, the existence and the conditions of **deposits** or other financial guarantees to be paid or provided by the consumer at the request of the trader.
- where applicable, the possibility of having recourse to an **out-of-court complaint and redress mechanism**, to which the trader is subject, and the methods for having access to it.

5.2.5 Consumer contracts

Where one party to the contract is in business and the other party is dealing as a private individual, this is known as a consumer contract and is subject to a range of legislation based on the premise that the consumer is in a weaker position, relatively speaking. The key legislation which implies terms into such contracts is the Consumer Rights Act 2015. Chapter 4 of the Act deals with the supply of services and stipulates that they must be carried out with reasonable care and skill (section 49). There are other related terms and a new provision introduced with the 2015 legislation was the right of the consumer to repeat performance if the original standard of the work had been unsatisfactory (section 53(3)).

5.2.6 Business-to-business contracts

Where both parties are in business, i.e. it is a business-to-business contract, then the law which preceded the Consumer Rights Act 2015 remains in force, i.e. the Supply of Goods and Services Act 1982 and the Unfair Contract Terms Act 1977. The former stipulates that work is to be carried out with reasonable care and skill (section 15). The Unfair Contract Terms Act 1977 essentially permits business contractors to negotiate their own terms, given that they are (theoretically) of more equal bargaining power than in a business-to-consumer contract.

For both consumer and business-to-business contracts it is *not* permitted to exclude liability for negligence which results in personal injury or death (Consumer Rights Act 2015, Schedule 2, Part 1 and Unfair Contract Terms Act 1977, section 2(1)).

5.2.7 Construction

Construction contracts are normally based on standard form agreements such as that produced by the Joint Contracts Tribunal or other standard form building contracts, for example, the NEC (originally New Engineering Contract) Engineering and Construction Contract. These contracts will obviously include the parties' obligations as agreed, costings, timings and, as mentioned above, will normally include provisions regarding dispute resolution. There is a range of standard form contracts, depending on the size of the project, and also contracts available which are suitable for consumers to use as a template when engaging their own contractors.

5.3 Tort

Tort is not a word used much outside more formal legal discussions. It is a collective word for a range of civil wrongs (other than contract) which cause loss or damage to the claimant (whether financial, physical or psychological), for which the defendant will be held liable. Areas of tort include negligence, private nuisance, trespass and defamation (libel and slander).

5.3.1 Negligence

It may be surprising to some to find that there are requirements at law which mean that standards clearly short of 'professional' may result in no finding of negligence (see *Barnett v Chelsea & Kensington Hospital Management Committee* (1968)), or that a particular claimant has no cause of action and, conversely, practice which may be considered 'standard' can be found to be negligent (see *Izzard v Field Palmer* (1999)). To establish negligence at law requires three elements to be fulfilled:

- Is a duty of care owed by the defendant to the claimant?
- Has that duty of care been breached?
- Is there resultant damage of a broadly foreseeable nature?

5.3.2 Is a duty of care owed for purely financial loss?

Liability for lack of an objective measure of care had been ascribed certainly from the fourteenth century (see *Bukton v Tounesende, The Humber Ferry case* (1348) on a ferryman's liability for a horse lost overboard), although not always under the banner of 'negligence'. It was applied where an underlying duty was understood by reason of the situation, such as passenger carriage, employment and third party loss for lack of care in business operations. The principle of liability for lack of care in the absence of an existing (notably, contractual) relationship for intrinsically *dangerous* items was extended in *Heaven v Pender* (1883), with the reasoning, essentially, that if you produce something, then you should take care that it is fit for purpose. This was picked up in one of the earliest negligent valuation cases: *Cann & Sons v Willson* (1888). A valuer was held to be liable for a negligent mortgage valuation when Mr Justice Chitty ruled:

> It seems to me that the defendants knowingly placed themselves in that position (of having their expertise relied upon by the claimant) and, in point of law, incurred a duty ... to use reasonable care in the preparation of the document called a valuation.

This may seem eminently sound, but it was soon overruled in *Le Lievre v Gould* (1893) in a case involving an inaccurate building survey on property in Ilfracombe. Taking the reasoning of an important House of Lords case on the production of an inaccurate company prospectus, *Derry v Peek* (1889), there was held to be no liability for false misstatement in the absence of fraud. Cases such as *Heaven v Pender* were distinguished as involving responsibility for an item which could cause physical damage if there was lack of care. Such cases *did* give rise to a duty of care, whereas negligent misstatement resulting in purely financial losses did not. But the law has since developed such that a duty *can* be owed for negligent misstatement resulting in pure economic loss where there is what is termed a 'special relationship'.

Liability for negligent misstatement substantially altered in 1963 with *Hedley Byrne & Co. v Heller & Partners Ltd.* This case established that there could be liability for pure financial loss (as opposed to financial loss consequent upon physical damage to person or property), in the absence of contract or even a direct relationship. The case took the dissenting reasoning of Lord Denning in *Candler v Crane Christmas & Co.* (1951), where the majority found that accountants were *not* liable per *Le Lievre v Gould*. Lord Denning presented a masterful analysis of authorities supporting the principle of a duty of care in professional cases, and clear reasoning as to why cases such as *Le Lievre* were wrong, both in law and morality. He took the now familiar position that it is only just to ascribe a duty of care when a professional knows their work, their statement is being relied upon. This dissenting position was picked up in *Hedley Byrne* which clearly set out that if a party knows that their statements are being relied upon, that there is sufficient proximity between the parties, then this 'special relationship' gives rise to liability, regardless of the existence of a contract.

Summary: a duty of care will be owed for financial loss where there is a 'special relationship', that is a measure of proximity between the parties whereby there is an assumption of responsibility. Note that even where a party has clearly been at fault, it may be that he or she owes a duty to party A but not party B, so now we consider to whom the duty of care is owed.

5.3.3 To whom is a duty of care owed?

In *Donoghue v Stevenson* (1932) Lord Atkin set out the now famous 'neighbour' test to the effect that a duty of care is owed to anyone who should reasonably be in mind as being affected by the acts or omissions in question. The scope of this duty was constrained in *Caparo Industries plc v Dickman and others* (1990), but the principle did not change, simply the level of proximity required to establish liability (for policy reasons running in a direct line to US Justice Benjamin Cardozo's famous warning in *Ultramares Corp. v Touche* (1932) about the dangers of ascribing 'liability in an indeterminate amount for an indeterminate time to an indeterminate class'). *Caparo* indicated that both the specific claimant and the purpose for which the information in question was being used need to be known to the defendant, *and* that to ascribe liability would be fair, just and reasonable. Note the recent Supreme Court review of *Caparo* in *Robinson v Chief Constable of West Yorkshire Police* (2018) where the commonly, but erroneously, held '*Caparo* test' approach of foreseeability and reasonableness was soundly discounted as a 'test' with the requirement to review each case on its facts. It was, of course, noted in *Caparo* itself that the reasoning should not be used as a 'test', highlighting how even apparently clear points from cases should not be treated as, effectively, statutory provisions.

In a valuation context, then, a valuer clearly owes a duty of care to anyone with whom he/she has a contract, i.e. the client. If there is a contract, then that would normally provide grounds for legal action where a valuer is suspected of falling below expected standards and it is unlikely, though not impossible, that a party with contractual rights would sue in tort.

Who else, in the absence of contractual rights, is owed a duty of care?

Valuation for commercial bodies – there tends to be no duty owed beyond the contractual as where a valuer is working for, say, a lender, then the borrower is expected to engage their own advisers who will, of course, have both contractual and tortious responsibilities (*Wilson v Messrs D M Hall & Sons* (2004)).

However, with lower-value residential properties, the lender's valuer knows that their expertise is likely to be relied on by the prospective purchaser and that it is unlikely that they will commission their own valuation. The courts have, therefore, ascribed liability to a third party, in the absence of contract, for lower-value residential properties (*Harris v Wyre Forest District Council* and *Smith v Eric S Bush* (1990)), extended this to modest commercial valuations (*Qureshi v Liassides* (1994) unreported) and noted the position in the RICS Valuation – Global Standards (UK appendix 10, residential mortgage valuation specification (1.2)). Note that where a lower-value residential property is acquired as a buy-to-let rather than owner occupation, then it is likely that there will no liability by the mortgage lender's valuer to the purchaser, with the expectation that the purchaser engages their own valuer, per *Scullion v Bank of Scotland plc* (2011).

Where there may be *no* duty of care

It should be noted that where there is no 'special relationship', then there can be no claim for pure economic loss in the absence of contract (per *Murphy v Brentwood District Council* (1991)). It has recently been attempted to circumvent this case with regard to allegedly negligent building inspection approvals, but the courts confirmed that there was no recourse in the absence of contract or a 'special relationship' (*Herons Court, Lessees and Management Company of v Heronslea Ltd* (2018) and *Zagora Management Ltd v Zurich Insurance plc* (2019)).

5.3.4 Breach of duty of care

The idea of an objective standard of reasonable care in the exercise of a skill was laid down in *Bolam v Friern* in 1957 (a case involving the electric shock treatment of a mental health patient but with reasoning applicable to all instances of those holding out particular expertise). The test is not whether another practitioner may have acted with more skill or may have come to a different answer. The test is, rather, whether the defendant did something which no other person of the same trade, profession or calling exercising 'ordinary' skill and competence could reasonably have done (*Bolam* at p. 21). This refers to the profession or skill the defendant is holding themselves out to have, regardless of their *actual* qualifications or areas of expertise. In a valuation context this comes down to the question of whether the value is one which no other competent valuer could have arrived at. Of course, this reference to an objective standard of 'reasonable care' and professional standards must be read alongside what the valuer in each case *indicated* that they would do, i.e. the terms of engagement.

Reference: Negligent valuation case summaries

In considering a practical application of what needs to be demonstrated to evidence 'ordinary' skill by a valuer, the reported cases highlight examples of valuer behaviours or omissions resulting in a judgment of negligence – see Foster and Lavers (1998), Murdoch (2002), and de Silva (updated annually) for case summaries.

Court decisions in this area of law clearly provide precedent and are thus beyond the merely illustrative, given the absence of legislation on negligence in the UK, but it is important that they are not to be trawled for simplistic rules as though they are some form of code. For example, court cases do not provide a 'law' that residential valuations should be within a 5 per cent margin. They indicate that this is the measure for 'standard' residential property valuations, subject to the facts of each particular case (see *Axa Equity and Law Home Loans Ltd v Goldsack & Freeman (1994), K/S Lincoln v CB Richard Ellis Hotels* (2010), *Blemain Finance Ltd v e.surv Ltd* (2012) and *Webb Resolutions v e.surv* (2012), the latter two cases being dealt with at the same time with *Webb Resolutions* being treated as the lead judgment). Valuations outside the margin guides do not signify automatic negligence. Rather, if a valuation is outside the margin, the burden of proof is on the valuer to justify his valuation which, on particular facts, he or she may succeed in doing.

5.3.5 Resultant damage

Most negligence actions are based on whether or not the defendant has exercised reasonable care (part two of the three stages listed above) or, less often, whether the defendant owes the particular claimant a duty of care (the first element of the three-stage test). It is relatively rare to be arguing whether or not the loss or damage in question is a result of breach. In a valuation context this element of the law may be raised in a situation such as that encountered in *BPE Solicitors v Hughes-Holland* (2017), where the loss was deemed to be as a result of the claimant's lack of professional judgment rather than the negligent advice.

Key good practice to evidence due care in valuation would include:

- Only valuing within area of expertise, with regard to type of property, geographical area, class of valuation, and any other relevant issue.
- Keeping up to date with all relevant legal, technical and professional matters.
- Taking adequate care and time (which will vary according to the nature of the valuation).
- Ensuring clear terms are set out and understood by the client, i.e. what is the precise nature of the working being undertaken. This will include appropriate disclaimers, such as are permissible in either business or consumer contracts.
- Sufficient site notes and appropriate final report.

5.3.6 *Measure of damages*

Where there is found to be a negligent valuation, the measure of damages is usually the difference between the 'true' valuation and the negligent valuation as assessed on the date of the valuation (*South Australia Asset Management Corp. v York Montague* (1996)). But where it is held that the valuer was providing *advice* on which the decision to buy was based (rather than simply *information*, i.e. a bare valuation), then the negligent valuer can be held liable for *all* losses which arise from the transaction, e.g. a due to a later fall in the market or property defects which were not apparent at the time (as explored in the *South Australia* case and *BPE Solicitors v Hughes-Holland* (2017) and applied in *Hart v Large* (2021)).

In *Hart v Large* (2021), a surveyor was found liable in negligence. Because this was judged to be an 'advisory' case rather than the simple provision of 'information', liability was for the full extent of loss resulting from the transaction. That this was advisory rather than information only was based on the sense that the transaction would not have gone ahead had the surveyor provided the full advice indicated as appropriate in the circumstances, in particular, that with a property with significant re-furbishment, a PCC (Professional Consultants Certificate) *must* be obtained. This is because there was no direct contractual protection as the property purchasers had not dealt with the builders and architects. A new build would, of course, have NHBC protection.

A valuation of £1.2 million was given on a residential property, with the 'true' value assessable on the day being £1 million. But there were hidden and emerging defects which could not have been assessed on the day which resulted in a judgment that demolition and rebuild was required with damages assessed at over £700,000. Leaving aside the differences of professional opinion on whether demolition and rebuild was necessary and, indeed, the figures put forward by the expert witnesses, some key practical points to emerge from this case include:

- The nature and constraints of report, i.e. the precise terms must be made clear to the client. Ensure they understand the limits of the work being undertaken.
- If an HBR (or equivalent) is commissioned, the surveyor should keep an ongoing review and recommend a full survey where appropriate.
- Where builders, architects or others are involved, the need for sign off and guarantees must be explained and stressed as vital, and the surveyor should not assume everything is likely to be alright, or that the solicitor will deal with it. Is a PCC (Professional Consultants Certificate), with the accompanying insurance, appropriate?
- Professionals should work closely together. Do not *assume* that lawyers will pick up all the issues, just as lawyers should not assume valuers and surveyors will pick up all the property issues.
- If something is not, or cannot, be viewed, then it should be indicated on the report (refer to relevant VPS, IVS and VPGA).

- Ensure insurance cover is at an appropriate level for the properties you are working on, bearing in mind that, although very rare, losses could be assessed at a full demolition and rebuild of the property.
- Emails and notes of telephone conversations should be filed. Telephone conversations should be followed up with an email to confirm what was said.
- Ensure emails and files are kept, and not lost when IT systems are changed.

For a detailed review of negligence as it applies to valuation, see the paper 'Negligent valuation de-constructed: What is negligence at law? What are the practicalities in helping to avoid a claim?' in the *Journal of Building Survey, Appraisal & Valuation* (de Silva, 2018).

6 The mortgage valuation

6.1 An overview

As discussed in Chapter 3, market forces ultimately determine price, and it is the role of the surveyor to interpret which forces will govern their ultimate valuation figure. How this is done depends on the client because different clients will have different drivers. Numerically, the two most featured clients to a transaction of residential property are the lender and the buyer, who both have an interest in its future saleability.

The experienced surveyor may well intuitively produce a figure for a property that fairly reflects market value. However, in the background is a complex process of comparison that is done almost subconsciously. The basis of that analysis is explained in Chapter 8 but valuing intuitively can also be accounted for through the process of expertise acquisition. This section will be of particular relevance to newly qualified surveyors, but it is hoped that even those more experienced may find this approach a stimulus to review their existing practices and reflect whether they meet current-day needs!

6.1.1 The lender's view

In recent years there has been a fundamental change in how lenders view property as a security for lending purposes within the UK. For many years, building society legislation governed the procedure that had to be adopted when valuing a property for mortgage purposes. This covered the majority of transactions and required a valuation for all properties, to ensure that they provided adequate security for the loan. Any factors that materially affected the valuation had to be reported. This still applies, but for those lenders that fall outside the legislation (e.g. banks), there has been an increase in regulation, with an emphasis on the customer's ability to repay the loan (affordability), meaning that whether the property forms a good security can be of secondary importance. This is especially true for the increasing number of overseas lenders who are more comfortable with this 'customer approach'. As a consequence, the emphasis on the valuation report has changed.

More recently there has been a huge increase in the use of automated valuation models (AVMs, see Chapter 10), desk-top and drive-past assessments, which are seen as quicker and cheaper for the lender. The use of these tools will only increase as their accuracy improves and the amount of property data that is made available increases.

As the residential market becomes more complex, and government regulation and intervention increase, lenders are also realising that not all valuers have skills in the

DOI: 10.1201/9780367816988-6

valuation of all property types. As a consequence, there has been a move towards the development of specialist valuation panels. This was initially restricted to high value properties, but has extended to valuations on leasehold properties, buy-to-let/HMOs (Houses in Multiple Occupation), and even new build properties. The modern valuer has to understand where their skills lie and look to develop these in order to meet the expectations of their client

6.1.2 The purchaser's view

The purchaser should, in theory, have a wider view than the lender, although this is not always the case. A significant proportion of customers assume that if the lender is satisfied with the property as security for a loan, then so should they be. The validity of this approach was set out in *Yianni v Edwin Evans* (1981) and confirmed in the case of *Smith v Eric S Bush* (1990) where Lord Griffiths stated, among other things, that:

> If the valuation is negligent and is relied upon, damage in the form of economic loss to the purchaser is obviously foreseeable. The necessary proximity arises from the surveyor's knowledge that the overwhelming probability is that the purchaser will rely upon his valuation, the evidence was that surveyors knew that approximately 90 per cent of purchasers did so, and the fact that the surveyor only obtains the work because the purchaser is willing to pay his fee.

This clear ruling has been slightly clouded, as the purchaser will not always pay a fee; some lenders do not make a charge directly, but the surveyor will still be paid. The debate will continue, as some would argue that if the cost of the valuation is included within the cost of the mortgage, the purchaser pays for it indirectly. The move by some lenders to non-disclosure of reports places purchasers in a further dilemma. Can they continue to rely on the fact that the lender is prepared to offer the loan? There may be a significant difference between the loan and the purchase price. This margin can accommodate a wide range of deficiencies in the property that may, in some cases, cost many thousands of pounds to rectify. Purchasers may be very disappointed with this gap between the lenders' criteria and their own expectations. Any disgruntled purchasers might be denied remedies in law. If this is the case, they will remain disillusioned with the professional who provided that report as well as that person's professional institution. To counteract this, professional organisations associated with the sale and purchase of property have long campaigned to educate the public. They urge people to obtain reports that more fully advise about the factors that will affect their decision to purchase. Such a report requires more detail than that provided by the lender's own valuer.

! **Benchmark: Duty of care**

Whilst in *Yianni v Edwin Evans* (1981) and *Smith v Eric S Bush* (1990) it was held that a surveyor owed a duty of care to a property purchaser with whom he or she had no direct, contractual relationship, this was found *not* to be the case for the purchase of higher valuer properties, commercial properties or buy-to-let residential where it is reasonable to suppose that the ultimate purchaser should get their own advice (and will therefore have contract rights) rather than rely on a mortgage lender's surveyor (see *Scullion v Bank of Scotland* (2011)).

An overview of the principles of negligence, including the circumstances of establishing that a duty of care is owed, is to be found in Chapter 5.

When undertaking a residential mortgage valuation, you have a duty of care to the applicant for the mortgage in respect of all matters that materially affect value, regardless of what the lender may require you to do.

6.2 Future issues

Do today's surveyors' reports still address the key points that are important to the lender? The lender assesses a loan on the basis of the borrower's ability to repay. However, there is also the fall-back of realising the money on loan by sale of the property taken in as a security, should this prove necessary.

The property is only a part of that equation, but a significant one, hence the reason for the variations in views by some lenders. Take the comparison with a company making a loan on a car. Lenders do not ask for a valuation of the car, so why should they with a property? It is clearly the purchaser of the car who needs to be sure that the car can perform as sold. The lender not only needs to know that the borrower can pay, but also that, in the event of default, they can realise the value of the asset to pay off the arrears accrued.

The problems of negative equity experienced in the early 1990s and after 2007 clearly identified a shortcoming within the existing valuation process. Predicting market movements or anticipating a change in price can prove to be very difficult. A lot of work has been undertaken on statistical analysis of the housing market both in North America and the Pacific Rim, but in this country the analysis has generally been retrospective. The ability to forecast accurately is an extremely complex process, especially when global influences on economies have to be taken into account. So, it is hardly surprising that the local valuation surveyor struggles with some of the modern aspirations of valuation. As the residential property market continues to develop, lenders may look to obtain advice on the potential future of house price movements, which will involve an element of prediction. Where this is the case, valuers must be careful to ensure that they have the tools, competence and PII (professional indemnity insurance) cover to provide such advice, as it brings with it enormous risks and the potential for misunderstanding, misrepresentation and confusion.

7 The RICS Red Book and Regulation

7.1 The Red Book

Valuation in the UK is generally completed by professionals who are members, fellows or associates of the Royal Institution of Chartered Surveyors (RICS); this is particularly a requirement of the lenders, who are obliged to ensure that valuations completed for them are undertaken by a suitably qualified professional.

RICS valuers are governed by standards laid down by both the International Valuation Standards Council (IVSC) and also by the RICS. These are the *International Valuation Standards* (IVS) and *Global Valuation: Professional Standards* (the Red Book) respectively. The standards laid down by IVS are high level and govern professional valuers across the world, including those allied to different professional bodies. The RICS Red Book is effectively the application of IVS for RICS valuers and comes with slightly more detail. Compliance with the key RICS standards is mandatory for members and RICS valuers need to be aware of these obligations in the operation of their daily business. Regardless, the Red Book is seen as the key standard to which all valuers are guided, and this is what they will be judged against through audits and, where things go wrong, in the courts or tribunals.

> **!** **Benchmark: International Valuation Standards (IVS) and the RICS Red Book**
>
> The high-level standards against which all RICS valuers are regulated and judged.

The Red Book was introduced in 1976 and was written as a *Guidance Note for the Valuation of Assets*, with the intention of bringing consistency and transparency to valuation practice. The aim was to develop confidence in the valuation product amongst lenders, investors and clients. Today, the Red Book is well established as a benchmark and aims to 'provide an effective framework within the rules of conduct, so that the users of valuation services can have confidence in the valuation of an RICS member'. In other words, it ensures the delivery of high-quality valuation advice.

At the time of writing, the Global Red Book is structured such that the two critical and mandatory requirements are detailed at the front in PS1 (requirement to comply with the standards) and PS2 (ethics, competency, objectivity and disclosures). Both

DOI: 10.1201/9780367816988-7

these Professional Statements underpin other Red Book requirements and apply to *all* valuations, be they Red Book or not.

The next section is a series of mandatory technical and performance standards, labelled VPS (Valuation Technical and Performance Standards) 1–5. These cover the practicalities of the valuation process, such as the contents of the terms of engagement, the inspection process, the compilation and writing of reports, bases of value and valuation approaches. All valuers must be entirely familiar with these, and with any subsequent updates in future iterations of the Red Book.

Finally, there are 10 global practice guidance applications, referenced as VPGAs (Valuation Practice Guidance – Applications). These are published as being advisory; however, in practice, any valuer who is working to the Red Book would be foolish to ignore the contents of these sections, as they lay down industry best practice against which work will be judged in the event of a valuation being challenged. The VPGAs cover the valuation process for particular asset types or purposes. Of particular interest to residential valuers are the sections that cover valuations for secured lending and matters that give rise to valuation uncertainty.

Having dealt with the Global Red Book, of additional interest to the residential valuer is the UK Red Book supplement. This document is published separately and covers the application of the Red Book standards in the UK market. Within this supplement is a lender-agreed specification for the valuation of residential property for secured lending. This is, in effect, part of the client brief and is a common specification to all UK residential mortgage lenders. It is a clear responsibility for all valuers in this market to be fully conversant with these requirements and to apply them. The requirements importantly lay down the bases of valuation and standard Assumptions and Special Assumptions to be applied in the mortgage valuation process.

7.1.1 Bases of value for mortgage valuations (VPS 4 and UK VPGA 11.2)

There are two bases of value detailed for valuations in the residential secured lending sector. These are:

> **Market Value (MV):** The estimated amount for which an asset or liability should exchange on the valuation date between a willing buyer and a willing seller in an arm's length transaction, after proper marketing, and where the parties had each acted knowledgeably, prudently and without compulsion.
>
> **Projected Market Value (PMV):** The estimated amount for which an asset is expected to exchange at a date, after the valuation date and specified by the valuer, between a willing buyer and a willing seller, in an arm's length transaction after proper marketing and where the parties had each acted knowledgeably, prudently and without compulsion.

! Benchmark: Bases of valuation

The defined bases of valuation provide a standard definition against which all valuations are completed.

The **Market Value** (MV) is the cornerstone of most residential valuations and valuers who apply it should take particular note of the willing buyer/willing seller assumption and also the fact that the transaction must be subject to proper marketing and with parties that are possessed of the facts and are not obliged to sell. This means that comparable evidence from sales that are subject to factors outside these requirements (such as repossession sales and sales to a 'special purchaser') must be disregarded. For secured lending purposes, the MV must take account of the property condition and reflect any factors that have a material impact on value. This means that, whilst the valuation should reflect the general state of repair of the property, a lender will not thank a valuer for deducting the costs of any general repairs pound for pound, or making a small adjustment to an otherwise acceptable purchase price to reflect minor condition factors. Where major repair issues or defects are identified, the lender policy will advise how these should be handled. This is often the withholding of a valuation until the full extent of the defects are understood. However, it is important for the valuer to remember that although the lender guidance is important, this must not override the duty of care owed to the applicant to the mortgage; sometimes this can conflict.

One of the key assumptions within the Market Value definition is that the parties have acted knowledgeably, but when the price is agreed, it is most unlikely to have been surveyed or valued by a professional surveyor at that early stage. This aspect coming later into the process can raise issues where factors are found that can materially affect value.

The **Projected Market Value** (PMV) (UK VPGA 11.2) is similar in definition to MV. The original intention of the PMV was a desire by lenders to have an indication of what was likely to happen for a property they had taken into possession. As such it is a 'special' basis of value, which assumes the property is unoccupied (and the adverse impact of this), it modifies the 'without compulsion' requirement for MV and acknowledges that the lender still has an obligation to achieve the best price for the property. However, by their nature, projected values rely wholly on assumptions, which may include some significant special assumptions. For example, the valuer may make various assumptions about the state of the market in the future, including yields, rental growth, interest rates, etc., which must be supported by credible studies or economic outlook forecasts.

Valuations to PMV are based on the following factors:

- the date is specified by the valuer (in fact, this rarely happens);
- it is a future exchange, based on what valuer knows today;
- it assumes completion and exchange on the same day;
- it is a simple numeric indication of short-term market trends;
- it gives an indication of whether the market will rise, remain static or fall.

7.1.2 Assumptions and Special Assumptions (VPS4, UK VPGA 11.3)

Having mentioned Assumptions and Special Assumptions above, it is worth touching on this subject separately, as both these elements have a specific definition in the Red Book, and it is important to understand the distinction between them. They are defined as follows:

> **Assumption**: 'A supposition taken to be true. It involves facts, conditions or situations affecting the subject of, or approach to, a valuation that, by agreement, do not need to be verified by the valuer as part of the valuation process.

Typically, an *assumption* is made where specific investigation by the valuer is not required in order to prove that something is true.'

Special Assumption: A *Special Assumption* is made by the valuer where an assumption either assumes facts that differ from those existing at the valuation date or that would not be made by a typical market participant in a transaction on that valuation date.

Where *Special Assumptions* are necessary in order to provide the client with the valuation required, these must be expressly agreed and confirmed in writing to the client before the report is issued.

Special Assumptions may only be made if they can reasonably be regarded as realistic, relevant and valid for the particular circumstances of the valuation.

!   Benchmark: Assumptions and Special Assumptions

These must be defined and documented before undertaking a valuation. Some standard Assumptions are outlined in the Red Book and are specifically accepted for residential mortgage valuations only.

It is clear from this that where a factor is unknown, it is sometimes reasonable to make an assumption on which to base your valuation. Many of the standard Assumptions that a valuer is entitled to use are laid down in the UK supplement to the Red Book and these include things such as assuming that a leasehold property is not subject to any unreasonable terms, or assuming an extension benefits from planning permission and building regulation approval (where there is no evidence to the contrary).

On the other hand, there are occasions when a client specifically asks the valuer to ignore something which is a plain truth for their valuation. The most common example of this is with new build properties where construction is incomplete (or maybe has not even started). Sometimes a valuer is asked to provide a valuation on the Special Assumption that the property is completed in line with the plans, which it most clearly is not.

The main thing to remember with *Assumptions* and *Special Assumptions* is that they must be agreed with the client 'up front' and must be documented in the terms of engagement and the report. Those Assumptions detailed in the lender protocol of the UK supplement to the Red Book can be taken as read for valuations for secured lending purposes, however, it is always important to ensure that the Assumptions are tailored to the property. In using assumptions and special assumptions, it is also worth being aware that, the more the valuation is based on elements that are not proven or true, the less reliable the final valuation outcome becomes.

7.1.3 *Valuer Registration (VR)*

In addition to their professional standards, the RICS also regulates all valuers through their valuer registration scheme (VRS). This scheme was set up in 2011 and membership is mandatory for all surveyors completing valuations under the Red Book. There is a cost to membership of VR and surveyors who are regulated by it must submit themselves to scrutiny over their compliance with RICS Professional Standards, as they relate to valuation. In practice, a VR audit is not a regular event, but the RICS adopt a risk-based approach to scheduling their visits and they can take place at short notice. The stance adopted by the auditors is one of support and assistance, particularly where member firms are open to their help. However, by far the best approach always has to be one of familiarity and compliance with the current published standards.

Benchmark: The Valuer Registration Scheme (VRS)

It is mandatory for all RICS valuers completing Red Book valuations to be a member of this scheme.

7.1.4 Insurance and valuations

As a valuer, it is critical, indeed mandatory, that you carry professional indemnity insurance that is sufficient and appropriate to cover the risks presented by your professional work. The specifics around the cover required are laid down by the RICS and you must always ensure that you are familiar with your obligations in this regard. This section does not seek to repeat RICS requirements, but will briefly outline some of the elements that you should consider.

Benchmark: Professional Indemnity Insurance (PII)

A business insurance, for practitioners who give advice or provide a professional service to clients. It covers compensation claims in the event of a surveyor being sued by a client for making a mistake that leads to financial loss.

Your insurances should always be seen as a key risk management tool in your business. At its most basic, it is a protection for both you and your client in the rare and unfortunate event that things go wrong. There are a number of things that you need to consider in arranging your insurance:

*   Is the cover sufficient for the nature of the work you do?
*   Is the cover for each individual case, or is there an aggregation clause that renders it insufficient in the event of a repeat error?
*   Is the cover on a 'claims made' basis? In short, does it cover you for when the claim comes in, which could be a few years after you actually did the work?
*   Have you ensured that there is 'run-off' cover in the event that you cease trading? Claims may still come in after your company stops completing work and you need to ensure that you have the cover to deal with these.
*   What is the level of the excess and is it levied per claim?
*   Has the provider set a maximum that can be paid out in any year and is this reasonable?

One point worth noting is that the risks associated with valuation can be further addressed by the employment of a liability cap. This is a mechanism by which a ceiling on the amount that can be claimed against the valuer on any particular claim can be set. If you seek to use a liability cap, it is worth taking legal advice as, in order for these to be enforceable, they must first be reasonable in the context of the customer or client relationship (this test will be tougher to pass if you are dealing with a member of the public rather than another business), and they must be fully documented in the contract or terms of engagement that you agree with your client.

Having covered the basics of insurance, it is also worth mentioning that, in the event of a claim and in the unfortunate circumstances that you find such a claim going to court, you will find yourself being judged against the standard of skill and care that would be employed by a fellow surveyor. This standard will be set by the guidance published by the industry and your professional body, as well as the reasonable standards that should be employed by any professional. If you can show due process and consideration in arriving at the contested conclusion, along with a full documentation of the process by which you came to your decision, you will find your case being viewed more favourably. It is for this reason we stress the importance of full site notes and documentation of the comparable evidence, adjustments and opinions that go into arriving at a final valuation conclusion.

8 The valuation process

8.1 The valuation appraisal process: overview

The appraisal process for residential property consists of a number of stages:

- collection of relevant information
- inspection of the property and its environment
- an interpretation of the extent and nature of the interest
- measurement of development potential or general enhancement of the property to maximise its benefits
- market analysis
- calculation of price/value/worth.

These may appear to be a complex series of questions that would take a significant amount of time to complete. In reality, for a mortgage valuation, it can take on average 30–45 minutes and for a Homebuyer Report 2–3 hours. The experienced surveyor may find it difficult to recognise the individual stages in the process, as so much of it will be done subconsciously. Also, in the majority of cases, the property will meet the normal expectations. In other words, fewer investigations will be needed, speeding up the appraisal process even further.

It is easy to believe that a surveyor receives an instruction, completes a valuation and this is the end of the matter. However, the process for undertaking a valuation is a lot more complex and is best approached by dealing with the process from end to end.

8.1.1 Pre-inspection research

Before you even visit a property, there is a lot you can, and should, do to ensure that you get the most out of the inspection, conduct your visit well equipped with the information you need and avoid wasting time inspecting a property for which you cannot or should not provide a valuation.

So, to start from the beginning, the receipt of an instruction, Figure 8.1 shows the pre-inspection actions.

Instructions can come from any number of sources, both corporate and private clients. It is important at the outset to get your instruction in writing, as a record that it has taken place and to establish who your client is, as this will affect your approach to the job and how you deal with them.

DOI: 10.1201/9780367816988-8

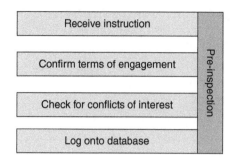

Figure 8.1 Pre-inspection groundwork

8.1.2 Purpose of valuation

You need to fully understand the reason for the valuation. For a private client they may wish to understand if a purchase they wish to make is good value for money, whilst a lender, for example, is just interested in whether the property is a good security for a loan and if it fits with their declared lending policy. These two example clients will come from a very different perspective and also a different level of understanding of the property market and thus should be handled distinctly and accordingly.

8.1.3 Express requirements

In analysing the instruction, you also need to understand any particular requirements that the customer has specified. This may be a need for the valuation to exclude an element of the property, or to be conducted on the basis of a specified *Assumption* or *Special Assumption* (see Chapter 7 on assumptions). If these requirements are acceptable to you, they will need to be agreed and clearly identified in the terms of engagement and subsequent report.

8.1.4 Report format

Finally, you will need to understand the parties to the transaction and the format in which the report is required. Private client reports will need to be completed in line with the Global Red Book requirements, whilst your corporate customers may have a wider contract with you and, very likely, a requirement for a specific short report format.

In more complex cases, the customer may not understand valuer constraints, e.g. that a mortgage valuation has to exclude speculative development value. It is important that in such cases the valuer and client understand what is to be delivered so that expectations are met. This emphasises the need to do some preliminary work to establish exactly what it is that is required based on the instruction and the type of property.

8.1.5 Terms of engagement

Normally terms of engagement have to be confirmed with the client prior to undertaking the job and allowing for a 14-day cooling-off period. The contents of the terms of engagement are laid down in the Global Red Book and this structure must be strictly adhered to. The only circumstances where this can be relaxed is where a lender or corporate client:

- has guidelines;
- has signed a pre-agreed set of conditions;
- provides a comprehensive instruction that refers to the RICS Valuation – Professional Standards Residential Mortgage Specification;
- has their own report format.

Where terms of engagement are sent out, these terms must have been fully brought to the client's attention and appropriately documented by the time the valuation is concluded, and prior to the issue of the report. This is to ensure that the report does not contain any revision of the initial *terms of engagement* of which the client is unaware.

 Benchmark: Terms of Engagement

These must be documented and agreed prior to undertaking a valuation. The contents of the terms of engagement are set out in the Red Book.

8.1.6 Conflicts of interest

Moving on to the next step, for each instruction a full conflicts of interest check must take place. For RICS members and associates there is an overarching requirement to exercise independence and objectivity in all instructions, and the process for checking this must be fully documented in the site notes.

The RICS lay out detailed rules around conflicts of interest, with which you must be familiar, but, in brief, a check must be made to rule out any personal conflicts (such as friends or relations) and professional conflicts (such as acting on two sides of a transaction, having a personal interest in the transaction, or other professional relationships that could influence your judgement). Where a conflict is identified, the client must be advised and the instruction declined.

 Benchmark: Conflicts of Interest

A conflict of interest check is a mandatory requirement for all work completed by RICS members. The current RICS Professional Statement on managing conflicts must be referenced.

8.1.7 Log onto database

It is only when all of the above is completed that you will log the case onto whatever log or database you are using.

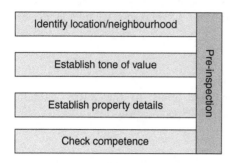

Figure 8.2 Pre-inspection research

8.2 Pre-inspection investigations

Figure 8.2 sets out the key headings within the pre-inspection research part of the process, once the case is logged onto the database:

8.2.1 Desk-top study

Having accepted an instruction, pre-inspection investigations should focus on understanding the nature of the property being valued and of the location in which it is situated.

This is the 'desk-top' study. This stage is very important for new surveyors and those who are working in a new geographical area. Once the address of the property is known, there are a number of quick checks that can produce some useful information that may have a material effect on value.

The pre-inspection checks can include:

- asking colleagues if they are familiar with the area or the type of property;
- referring to historical and current survey maps for the area. For several urban areas, maps from the mid-1800s have been published commercially. A few minutes spent looking at these can give a quick insight into the history of the site. This might be specifically useful in respect of building movement and contaminated land issues as the position of old quarries, pits, ponds, streams, etc. could be revealed.
- looking at reports produced by the surveying organisation from the files of similar properties in the area;
- check any general files or special databases that the office maintains or subscribes to;
- geological maps;
- radon maps;
- crime statistics;
- information on mobile phones;
- waste disposal;
- mining;
- fracking, although this may cease as a requirement dependent upon government determination;

- socio-economic criteria;
- school standards;
- identifying outbreaks of invasive plants;
- valuation of comparable sites, there a number of these available on the internet, some requiring a subscription, but they provide useful photographic and sold price information of comparable properties.

8.2.2 Neighbourhood

As an extension of location, it is also worth considering the concept of a 'neighbourhood'. This is something that becomes instinctive to the experienced valuer, but is critical to understand for those just starting out.

A neighbourhood can be defined as 'the environment in which the property is located that has a direct and immediate impact on its value'. That is, a defined area that is subject to the same (or similar) market forces.

Before undertaking a valuation exercise it is important to establish an area from which comparable properties can be sourced. This will be a geographic area that shares similar characteristics, as defined for market analysis or modelling purposes. A neighbourhood could have relatively few properties, or could extend to a sizeable number of units from which it is sensible to select properties that can be used for comparable evidence.

In defining a neighbourhood, the valuer should look at the quality of the area, the proximity of facilities (such as a top-rated school), the geology of the site, infrastructure (such as roads, railways, etc.), socio-economic factors, occupation types (owner occupier or rental) and, in some cases, the religious or ethnic origin of the occupiers. This is all based on the fact that people like to live with like-minded people, so in determining value, the occupiers in these areas will value or appreciate the same factors.

No evaluation of the neighbourhood would be complete without some consideration of the consumer. In many cases the surveyor may not even be aware of the actual purchaser and is more concerned with the lender. Generally trends develop where consumers of like minds and financial circumstances tend to be attracted to particular areas. This has resulted in a variety of consumer-based statistical surveys being undertaken by companies who produce consumer classification data, such as Mosaic (produced by Experian) or Acorn (produced by CACI Ltd). The underlying information for these databases is the census, although other types of information may support them. The approach is to evaluate socio-economic groupings based on postcode and the resulting classifications are useful to companies needing marketing advice to target areas for specific product sales. Essentially these surveys will give an indication of the typical lifestyle of someone living in a certain area. Further investigation can reveal what products that particular lifestyle will want to buy, so companies can target them with sales campaigns accordingly. These classification companies are well positioned to expand this data to provide value comparisons. However, a slow drive around the area before undertaking the actual inspection is a good way to establish a lot about an area.

Benchmark: Neighbourhood

The benchmark here is the **neighbourhood** and the question may be whether one actually exists. In an urban area it is almost always the case, but in a rural area, then the neighbourhood is the countryside which may have very few properties, especially those that have sold recently.

8.2.3 Tone of value

Having determined the 'neighbourhood' from which the comparable properties should be selected, it is then possible to establish the 'tone of value' for that neighbourhood.

In understanding what 'tone of value' is, it is worth looking at its origins. The expression comes from the world of rating valuation, where reference was made to the 'tone of the list'. In order to establish this tone, the rating valuer would look at a street to see what other rating assessments had been set or agreed and, based on this, look to establish where the subject property fitted in the list.

If we bring this across to the residential valuation market, we are looking to understand what the tones of value are within the area that contains comparable properties to the one that we are considering (or the neighbourhood we have identified). In this consideration we have to understand the range of properties in this neighbourhood, which could be different in detachment and style to the one that we want to value. It may be that we start off with a wide range of tone, but seek to narrow this down later when we understand more about the subject property. At this stage it is about understanding the nature of the area in which your property sits.

Benchmark: Tone of value

The benchmark here is the **tone of value** (range of values in a given area). It is within the tone of value for this location that the value of the subject property will sit. Is the price of the property at the top end of the tone, at the bottom, or somewhere in the middle? The value should only be at the top if it is the best property in the area. If not, then the valuer may be looking at adjusting the value by comparison to other properties in the locality.

8.2.4 Property details

As a further pre-inspection check, you need to understand all you can about the subject property. This is best done using the plethora of online tools that are available and will assist you in planning how much time you will need on site and any health and safety preparations you may need to make. It is very hard to keep one page ahead of the customer when so much information is now freely available. But knowing the information is one thing; being able to apply it in the context of value to property is another and that will be the surveyor's strength.

8.2.5 Property details – legal issues

At this point it is also worth establishing as many legal facts about the property as you can. These will not always be present in the instruction but can have a material impact on the value and/or marketability of a property. If the property is leasehold, the remaining lease term is critical, along with any liabilities that come with the lease such as ground rents and potentially onerous clauses. In addition, if the property has been altered, is there planning and building regulation approval, is the property subject to any legal restrictions or covenants? Is it listed? And, if a buy-to-let, does it require licensing? All these elements can play a critical part in valuation and information established up front can often be clarified with an occupier at the time of the inspection.

8.2.6 Competence

Finally, it is worth pointing out that if, at any time, you establish that the valuation of a property is outside the scope of your experience or competence, the instruction must be declined. Do not attempt a valuation for which you do not have the skills or experience. As soon as you realise this, the client must be informed, and the instruction declined whatever stage of the process you have reached.

It is important that all the information gleaned from this research is recorded, and this should form part of your site notes, which will be covered in more detail in Section 8.3.

8.3 On site

8.3.1 Location

On arriving on site, armed with pre-research, you will have a good idea what you are expecting. The process is outlined in Figure 8.3. It is, however, always important to have a look around the area in the vicinity of the property to understand any 'on the ground' issues that could impact value, as well as a look for 'for sale' and 'sold' boards, indicating potential comparable evidence.

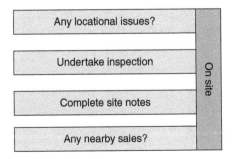

Figure 8.3 On-site inspection

The Global Red Book states that inspections and investigations must always be carried out to the extent necessary to produce a valuation that is professionally adequate for its purpose. The valuer must take reasonable steps to verify the information relied on in the preparation of the valuation and, if not already agreed, clarify with the client any necessary assumptions that will be relied on.

These general principles are supplemented by the following additional requirements embodied in VPS 1 and VPS 3:

- Any limitations or restrictions on the inspection, inquiry and analysis for the purpose of the valuation assignment must be identified and recorded in the terms of engagement and also in the report.
- If the relevant information is not available because the conditions of the assignment restrict the investigation, then if the assignment is accepted, these restrictions and any necessary assumptions or special assumptions made as a result of the restriction must be identified and recorded in the terms of engagement and in the report.

The guidance in the Red Book in box 2.6 is the generic requirement for any instruction, but a lender may have a different approach.

8.3.2 Inspection

It is important to understand the purpose of the inspection and, if undertaking a valuation, rather than a full survey, to adjust the depth of the analysis of the property accordingly. This chapter will look at how these requirements can be differentiated.

The Red Book states:

- A valuer will be familiar with, if not expert on, many of the matters affecting either the type of asset including, where applicable, the locality.
- Where an issue, or potential issue, that could affect value is within the valuer's knowledge or is evident from an *inspection* or examination of the asset, including where applicable the immediate locality, or from routine enquiries, it should be drawn to the client's attention no later than when the report is issued, and ideally in advance of the report in cases where the impact is significant.

Precise descriptions are given in the RICS's *Red Book* in respect of valuations and in the *RICS Home Survey Standard* (RICS, 2019b) for surveys.

- The '**mortgage valuation**' is a report for a lending institution subject to agreed limitations for the expediency of the house buying process, where a mortgage to purchase is required. Given that the client is a commercial organisation, and that speed is of the essence in processing the valuation, it has been accepted that a limited inspection will take place to try and identify all matters that materially affect value (Building Societies Act 1986, section 13).

KEY POINT

Not all valuations are undertaken for building societies, however, the Building Societies Act 1986 provides the legal basis for many mortgage valuations and most of the subsequent lender requirements as set by the Financial Services Regulatory bodies follow this principle. It is also the terminology used in the Red Book for secured residential valuations.

Due to the limitations and the lack of full investigation, the valuer can make a number of assumptions that those limitations would not have revealed anything that would materially affect the value; whereas:

- a '**survey**' is a report based on a lengthier and more in-depth inspection advising primarily on 'condition', but which can, and very often does, impact on value. Surveys can also be subject to practical limitations and may also involve assumptions, but these are to a lesser degree.

In general terms, the scope of a mortgage valuation or equivalent inspection is less than a survey but there is a requirement for the advice to cover 'the nature of the property and factors revealed during the inspection that are likely to materially affect its value'. A duty of care applies to both the lender and the applicant to a mortgage and that duty will vary depending upon the individual buyer as defined in *Smith v Eric S Bush* (1990) (see Chapter 5). Some limitations are outlined in the Red Book and a more comprehensive list of assumptions are defined in the UK Supplement (VPGA11), where the valuer can limit the scope of what is required. Although these are key differentiators, it is up to the discretion of the valuer to determine the limit of the inspection and where suspicion of an issue is identified, then the valuer has to determine what further action is required to ensure that a 'trail' (*Roberts v J Hampson & Co.* (1989)) is followed to establish if it is a factor that materially affects value.

We make this fundamental distinction because there has long been, and continues to be, confusion amongst consumers and the market about the differences between the two types of report. Thus, negotiators in estate agency offices and mortgage brokers still regularly, in the writer's experience, refer to the mortgage valuation as a 'survey'.

The term 'surveyor' has been adopted and applies to those professional people who see themselves as either 'valuers' or 'surveyors'.

This chapter looks at how to carry out an inspection from first principles with the focus on valuation inspections. For many experienced practitioners this may appear too basic, but it has been included for two reasons:

1. the book is designed for a broad readership, including cognate and non-cognate entrants into the profession. For these readers introductory material is necessary.
2. for experienced professionals it will be useful to 'check' current practice against what is deemed to be acceptable practice.

Any successful inspection will consist of three stages:

1. **preparatory work**– discussed previously in this chapter;
2. **carrying out the actual inspection itself;**
3. **reflection of what has been seen before preparing a report on the outcomes**.

Preparation of the report is addressed in Chapter 9.

The overriding requirement is that your valuation must reflect all factors that may materially impact value. This is potentially a very wide-ranging remit, but is limited by the scope of the inspection and the report as specified in the UK Supplement to the Red Book, with which you will need to be very familiar.

In addition, if you are completing an inspection for secured lending purposes, you will also need to be familiar with the current lender protocol in the UK Supplement and also the specific requirements of your client. For example, some will require you to undertake a roof space inspection and others will not.

8.3.3 Site notes

What cannot be stressed enough is the importance of keeping a comprehensive record of your investigations, commonly known as 'site notes', taken throughout the valuation process. Whilst these are not necessarily gleaned physically on site, they are your evidence of what you saw on the day, from investigations both before, during and after your inspection, and they can be in hard format or recorded digitally. A comprehensive set of site notes will enable you to respond to post-valuation questions (PVQs), which can arrive some weeks later (and after you have seen many other properties in the interim). In addition, they are your evidence in the event of a complaint or challenge and it took a series of court cases going back to the 1980s and the 1990s to highlight exactly what was required in not only the production of comprehensive site notes, but other aspects of practice (see Chapter 5).

Depending on personal style and preference, surveyors will either make written notes and sketches to record their impressions of a property, use a form of computer tablet or dictate their thoughts into a personal recorder. Opinion is divided on the most suitable method and is discussed in the case of *Watts v Morrow* (1991), but this did not include computer equipment as such technology was not available then. The principles set down in that case are, however, important in ensuring a comprehensive record of all your investigations. Although this case related to a structural survey, the fact that the surveyor used a personal recorder was a matter of concern to the judge. The particular worry was that the surveyor dictated his report directly into the machine not just his site notes. Once back at the office, the tapes were passed straight to the secretary who typed them up. The report was amended and then sent to the client. This resulted in the report being 'strong on immediate detail, and I regret to have to say, negligently weak on reflective thought', according to the judge.

This ability to ponder over a recent survey is an important part of the process. To be able to look at the dwelling as a whole is vital. Seeing how the different elements interact and draw reasoned conclusions is what surveyors are paid for. The absence of any site notes makes this much more difficult. In the *Watts* case, this lack of written notes also meant that the surveyor had nothing to highlight a 'trail of suspicion'. He had trouble recalling some aspects of the property which was put down to not having a written record of the survey. This attitude is supported in the survey standard (RICS 2019b), requiring the production of an accurate and comprehensive record of the property at the time of inspection to allow reflection before the service is delivered.

In *Bere v Slades* (1989), a valuer admitted that he had no recollection of a valuation he had carried out other than from his site notes. The court decided in favour of the valuer. This was probably helped by the positive impression of competence promoted by his good record keeping.

The surveyor's comprehensive contemporaneous site notes were also a key feature in the finding in favour of the surveyor in *Hubbard v Bank of Scotland* (2014) in the Court of Appeal.

The importance of making sketches was mentioned in the judgment of *Fryer v Bunney* (1982). Here the surveyor had checked a property with a moisture meter and reported that no dampness readings had been registered. In the event, there was indeed dampness that could have been detected with the moisture meter had it been used more effectively. Murdoch and Murrells (1995, p. 94) concluded that: 'The moral of this story must be for surveyors to prepare a very simple sketch plan of a property as it is surveyed, marking roughly where a Protimeter has been applied.' (Other kinds of moisture meters are now available.)

Any potential court case can be lost or won on the quality of site notes (see *Ryb v Conways Chartered Surveyors* (2019) where a lack of site notes and photographs was unhelpful to the surveyor in a claim finding him negligent for not spotting Japanese knotweed).

In conclusion, if there are no specific organisational quality assurance procedures, the following record of a survey should be made:

- clear handwritten notes taken on site with enough sub-headings to indicate which parts of the building the notes relate to. Sketches can add to the descriptive value of the record.
- the notes should be legible and set out in a logical order. Ideally, they should be in pencil in wet weather and re-written in the dry, if necessary.
- if a personal recorder is used, then the tapes should be transcribed as soon as possible. The transcribed notes should be used when compiling the report.
- a computer tablet should be able to replicate written site notes;
- photographs should be taken to illustrate the notes;
- any additional information from either the pre- or post-inspection routine should be included in the record of inspection. This includes the details of conversations with your client and any agents you consult.

The site notes should be filed along with the rest of the client information for as long as practicably possible. They can be in either hard copy format or scanned as a computer record, but in all cases, they should be secure from being hacked and destruction by fire or flood.

! **Benchmark: Record of your investigations**

The benchmark here is a **record of your investigations**. It is from this record that everything about the work you are doing is sourced, so ensure they meet the RICS guidelines, but that they also suit your working routine and help you do your job effectively

As a part of completing those site notes, make sure you take some time to pause at the point of entry, usually a gate or a road frontage. Look holistically at what you are going to inspect. Not only the property itself, but how it sits on the plot. Think of that property as a box and ask yourself a few questions:

- Does it have all the sides, are they complete, do panels make up part of the structure?
- Where are the apertures in that box, i.e. the windows and doors? Do they weaken the box?
- What is the material making up that box?
- Are there any areas that are prone to water, whether it is flooding, exposure or leakage?
- What is sitting on top of the box? Is it a triangular roof or a flat one?
- Is there anything encroaching on the site, such as trees or rights of way?

At this stage you are beginning to build a picture and identify areas where there may be a trail that you want to follow within scope or refer for further investigation. Hopefully it is looking good and there are no significant issues, but your job is to find the unexpected.

8.3.4 Setting the scene with the occupier

In most cases the occupier will be the vendor. They may be your client, or may be a tenant of your client and depending on your remit, this may affect the questions you want to ask. However, whatever the remit, the first thing is to put the occupier at ease and explain what it is you intend to do. Have some form of identification and make sure your shoes are clean.

Check for any issues that may affect your inspection, such as dogs, children, sick residents or someone asleep from a night shift.

It is important to set out how you wish to undertake the inspection, what tools you may need to use and that you will need to take photographs, but ensure no residents are included in the photographs.

Other matters you may wish to establish:

- ask general details about the property, e.g. its type, size, age, etc. These are useful facts that can help inform the office-based preparations.
- inquire whether any work has been carried out to the property, such as extensions, alterations, repairs, etc. If yes, ask them to sort out any documentation they may have. This could include guarantees, planning and building regulation permissions, etc.
- check whether there are any bats in the property or conservation-related issues.

During the inspection, take great care not to damage the property in any way. Typical examples have included:

- marking wall and ceiling surfaces when assembling ladders for loft inspections, etc.;
- soiling carpets with muddy boots, dog faeces, etc.;

- disturbing decorations when unscrewing access hatches, using moisture meters, etc.;
- knocking ornaments off mantelpieces, shelves, etc.

If the house is empty, it might be appropriate to take a few photographs of the positioning of any expensive-looking ornaments on a time-stamped camera.

8.3.5 Inspection procedure

The method described below is just one possible way to complete an inspection. The main rule is that, whatever your approach is, it must be methodical and systematic, and the same routine should be followed in every inspection. This can help create an impression of competence if challenged in court. One of the first things to do is to identify any health and safety issues, these are specifically covered in Volume 2 of this book and in the RICS current edition of *Surveying Safely: Health and Safety Principles for Property Professionals*, and we cover a few rudimentary points later in this chapter.

A typical inspection could include the following.

Internal inspection

Work around the dwelling internally:

- inspect the loft space first (if there is one); for an MV inspection this would be head and shoulders only if the lender required it, currently it is not mandatory, although it is an area of the property that can reveal a lot;
- inspect the rooms on the uppermost floor by working around that entire floor in a clockwise direction;
- finish on the landing and inspect the stairs down to the next floor;
- follow the same process on each floor down to the lowest one;
- inspect any cellar or basement, sub-floor voids are for the more detailed level of survey.

In each room, inspect the various elements in the following sequence:

- ceilings;
- walls (including skirtings);
- floor;
- windows and doors;
- heating;
- electrics;
- plumbing;
- other amenities (e.g. toilets, basins, sinks, etc.);
- fittings and fixtures (e.g. cupboards, fitted wardrobes, fireplaces, etc.);
- any unusual or special features.

External inspection

It is a good idea to go outside towards the end of the survey, to avoid the problem of tramping muddy boots around the house.

The external inspection should include:

- the main elevations, including all secondary elements (doors, windows, etc.);
- observable roof surfaces, including chimneys stacks, etc. Flat roofs may be inspected from the windows of upper rooms. Rainwater goods should be assessed at this point.
- any significant garden features, such as retaining walls, etc.;
- all outbuildings;
- boundaries, fences and gates;
- any special features such as rights of way, etc.

The use of cameras on poles are becoming more significant tools, but are not mandatory for the mortgage valuation inspection at the time of publication.

Drainage inspections are not included in the mortgage valuation inspection.

This systematic approach is fine in theory, but some defects may show signs in more than just one space (Wilde, 1996, p. 18). Therefore, although you do not inspect the drains, it may be that above ground there is an undulation that may indicate some subsidence of the drain run, and this could be material. Therefore, there is a trail to follow, but this would be a referral due to the limitations of the inspection.

Once the inspection has been completed, always inform the owner that you are leaving.

8.3.6 Inspecting flats and maisonettes

The RICS Red Book, in the section on secured lending for residential properties, identifies what parts of a flat or a maisonette should be inspected. Although this may vary between organisations for lending purposes, the key features include:

Externally: the exterior of the whole property that is accessible from the common areas. This is to gauge the general state of repair and includes the normal rules governing flat roofs. The extent of this assessment was described by Holden (1998a, p. 26) who suggested that the surveyor must 'take reasonable steps to establish any specific problems which the legal advisors should raise with the management company'. Outbuildings should also be inspected in a superficial way apart from those used for leisure activities, which are mostly excluded for mortgage valuations.

Internally: the interior of the actual dwelling, the communal area from the front entrance door of the whole building to the flat entrance door and the remainder of the staircase. This will include any roof spaces that are accessible from within the actual property on a head and shoulder basis. The same rule also applies to cellars and basements.

8.3.7 Personal safety

Due to the solitary nature of the work, surveyors can be vulnerable to personal attack and injury. The best advice on how these risks can be minimised has come out of the work of the Suzy Lamplugh Trust. This is the national charity for personal safety and takes a positive approach to the aggression and violence that professionals may have to face. They have collaborated on a number of publications aimed at people who work alone in other people's homes.

8.3.8　How long should it take to do a valuation?

In the past, case law was helpful, but the key cases go back to 1989 and 1990 (*Roberts v J Hampson & Co* (1989) and *Lloyd v Butler* (1990) respectively). Circumstances have changed dramatically since then and most valuers would say that they spend as long doing the pre- and post-valuation preparation as they do undertaking the inspection itself, such is the need for wider considerations since more information has become available on the property. What still applies to some extent is what the judge, Ian Kennedy J, stated in the *Hampson* case:

> It is inherent in any standard fee work that some cases will colloquially be 'winners' and others 'losers', from the professional man's point of view. The fact that in an individual case he may need to spend two or three times as long as he would have expected, or as the fee structure would have contemplated, is something that he must accept. ... If, in a particular case, the proper valuation of a £19,000 house needs two hours work that is what the surveyor must devote to it.

The valuation of £19,000 dates the quote and it is much more frequent, now, for the valuer to specify the trail of suspicion and recommend a referral, so reducing the time on site. However, that would be no defence if there was a trail to follow and it was missed. This is where proper preparation can help.

8.3.9　Referrals

The inability to fully investigate the true nature of a defect during a standard inspection for a mortgage valuation is a key differentiator from the survey. The limitations on time to undertake the inspection and meet tight deadlines set by the lender mean that the valuer is entitled to defer a more detailed investigation that may occur with a survey. This does not mean an issue can be ignored or an assumption made that a defect will go away. Any issues such as indications of penetrating dampness, that usually will have a material impact on value if left to progress, need to be addressed. If the cause of the defect is immediately apparent and the valuer can estimate the likely cost and its impact on value, then this can be reflected in the current value. Where the cause cannot be seen and fully diagnosed, then a recommendation for further investigation may be necessary. This is where the valuer needs to make a decision as to whether an assumption can be made that this can be readily reflected in the value or whether a value cannot be produced until the further investigation has been completed. No buyer or lender will want any delay in progressing to a point where a mortgage offer can be released or that delays the exchange of contracts. However, it would be negligent if the valuer allowed something to progress that would materially affect the value and it could not be fully accounted for. Ignoring a defect helps no one, although the valuer may be put under pressure to do exactly that.

Understanding the pathology behind the defect continues to be a critical part of any inspection, no matter what level. This allows the valuer to decide which defects can progress with assumptions as to how they will impact the valuation, and those that need further investigation and are effectively a 'showstopper'.

In most cases the action will be to get a contractor to quote for repairs prior to exchange of contracts, as this fulfils the duty of care to both the lender and applicant to the mortgage, so they progress in an informed way.

8.4 Post-inspection comparables and reporting

8.4.1 Research comparables

With investigations and inspection complete, it is now time to turn to the process of establishing a value for the property by means of comparable transaction analysis. This is start of the post-inspection process, as shown in Figure 8.4.

This valuation technique relies on the concept of 'substitution', in other words,the acceptance that a knowledgeable and prudent person would not pay more for a property than the cost of acquiring an equally satisfactory substitute.

For this purpose, a substitute can be defined as other properties that are available:

* within a time-frame;
* in a suitable location;
* with similar utility and desirability.

In the context of valuation, 'utility' is a more generally a commercial term but does have some reference to residential property, especially if you are valuing for buy-to-let purposes. It is a function of the characteristics of the land, buildings and location. The price paid or rent offered for such properties is a reflection of the utility to the user, i.e. how convenient it is: proximity to work, shops, schools, etc., as all these have a cost of travel.

Beyond this, we have to then make an allowance for any variables in attributes between the properties. These variable attributes are all the factors that make up the interest that a buyer will have in the property.

The term 'attributes' can be defined as:

* characteristics;
* the qualities of a property and its immediate environment;
* its positive and negative features relating to the market in which the property is based;
* legal aspects that affect the benefits for the owner/occupier.

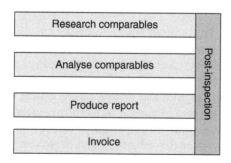

Figure 8.4 Post-physical inspection work

Additionally, within each of these, there are other important factors. For example, under 'type of property':

- size of property;
- the dwelling's attachment, e.g. semi-detached, detached, terrace, etc.;
- constructional features may be significant especially if non-standard.

The relative importance of each one of these attributes may vary considerably between different geographical areas:

- character of location can play an overriding role, regardless of type of property;
- school catchment areas may influence price, regardless of other characteristics;
- number of bedrooms is probably the key attribute within the accommodation section within the UK.

Most importantly, some of these features may change in priority over time and the valuer needs to be aware of such changes.

These days, a valuer has access to a number of comprehensive on-line comparable databases, giving an extensive record of property transactions in the locality. The amount of information given about each property is expanding all the time; though the surveyor must use their judgement on how much reliance they can place on elements of the information (such as floor area) without being fully aware of its source and reliability. The analysis of this information hinges on the interpretation of how those attributes combine. This is done by identifying those circumstances that differ from the norm (or the surveyor's expectations of it). In essence, the surveyor must be able to record the key attributes of comparable properties so that a full evaluation can take place. Knowing what information to collect is based on a sound understanding of what you are trying to achieve at the outset, i.e. comparison of the subject property with those that are:

- in a similar location
- with similar accommodation
- similarly designed
- in a similar condition
- with a similar quality of finish/specification
- having similar amenities
- in similar market conditions
- with a similar market exposure.

The best comparisons have established values. Therefore, the collected information is used to establish the differences between the subject property in its location and the best comparisons known to the surveyor that have been sold in the open market at a point in time.

8.4.2 Analysis of comparables

The next step is to determine how important those differences are.

The process does not start at the property but by considering whether anything has changed nationally, regionally or locally since the last inspection. This may include the density of 'For Sale' boards or, on a more macro scale, a change in interest rates.

So information collection starts by reading local and national news headlines and commentary.

Special factors may also impact upon the price. The information given to the surveyor at the outset may not include details of why the property is being bought. A simple example is where someone is moving to be close to a new job. Moving to be close to a family member is another example. This may result in the individual paying more for a particular property in order to be sure of securing a property in the desired the location. This may produce a figure in excess of the general market value and therefore may not be repeatable. Regrettably, this level of detail may not be apparent to surveyors as their only personal contact is often limited to the vendor.

It is also worth pointing out that the valuer does not always have access to the full information about a property. Where this is the case, it is the responsibility of the valuer to take reasonable steps to seek to establish the information required. Where this is not possible, they will need to fall back on *assumptions*, either those laid down in the Red Book, or any other assumption necessary to produce the instructed valuation (see Chapter 7, Section 7.1.2). Where an assumption is used, this must be made clear to the client in terms of engagement, or written communication, and must also be specified in the report.

8.4.3 *How to use the information*

Having determined the objectives for collecting the data, the next stage is to collect the key attributes of the property. It is important to record any issues that are unique to the property. The reasons for this can be listed as follows:

- An experienced surveyor will have in mind a comparison for the subject property. The similarity will be based upon the expertise of the surveyor and the quality of the existing records.
- The process will involve an objective and subjective ranking by referring to established scores for each of the key attributes. Generally this is done without consciously thinking. A few comments are then made in the site notes, usually where there are deviations from the norm.
- The collection of data adds to existing records and the knowledge base of the surveyor.
- In addition, it is important to establish those features within the property that require enhancement to bring the dwelling up to the expected standard for its type and age.
- Where the clients are known and the surveyor has had the chance to evaluate their needs, then an on-site assessment that measures the real or potential benefits to that client would also be significant.

The process can be summarised as:

- collection of data to use for comparison purposes;
- a ranking of the attributes and assessment of them by reference to existing comparables;
- identification of defects and enhancement potential;
- evaluation of the match between inherent benefits and client needs.

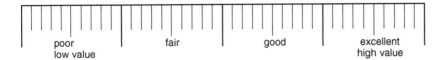

Figure 8.5 Scale of comparison for different properties. The key to successful valuation is correctly placing the property on a relative scale.

In essence, the surveyor has to determine where the subject property lies within the range of the comparables of which they have knowledge. So in the context of the scale shown in Figure 8.5, property comparisons may fill the whole range from poor to excellent, and more likely they will be nearer the average. It is for the surveyor to establish whether the subject property is better or worse. An added complication is when there are a variety of factors, some of which are better and some worse. In these instances, remember to keep it simple. Map out all those factors that are better and those that are the same or worse – imagine them on a scale like Figure 8.5. More help is given with this technique later.

8.4.4 Market analysis

In this section we highlight all the factors or attributes that make up the analysis of a property by comparison to the current state of the market. The most important aspect is the local market, but this should not ignore macro considerations. The general state of the economy must be accounted for, as well as indicators of trends within the housing market taken from leading parts of the country, such as London. For example, in order to bolster the economy in recent years, interest rates have been low for some time. However, political uncertainty in the run-up to Brexit resulted in a muted property market as people held off selling until an element of certainty returned. Brexit uncertainty overlapped with the global COVID-19 pandemic, further chilling the market. The extent and timing of such events are often difficult to anticipate. Local conditions, such as the closure of key businesses, may have an impact upon the local market, contrary to conditions at a national level. The ability of the workforce to find alternative employment without moving location is a key determinant of the impact. The publication of a regular analysis of the economy by the RICS is an indicator of the relevance of this topic and increasing expectations of the customer.

Generally, the surveyor undertakes the analysis of the market based on retrospective information. This poses significant problems for people who want to know how that value *will* perform, as this information does not forecast what will happen in the future.

In *Corisands Investments Ltd v Druce & Co.* (1978), the surveyor was found negligent for not allowing for the speculative element in the property market. The judge based his decision on what 'an ordinary competent surveyor' would have done. The test of reasonableness will usually apply.

The surveyor can anticipate some of what will happen to the property, especially with regard to condition. This provides a useful comparison between the effect of condition on value (tangible) and those factors of a more subjective nature such as micro and macro-economic forces and their influence on value (intangibles). Identifying faults in design or structural elements will lead the surveyor through a process, resulting in a recommendation to undertake repairs at a cost that can be considered in arriving at a market value. By referring to past experience of similar conditions, the risk to the property can be fully assessed and the surveyor can be fairly certain of the likely outcome and cost. It is different with the economy as a whole. The chain of events that can trigger a recession within the property market, either nationally or locally, can be very complicated. In addition, as each scenario pans out in a different way, with any number of influencing factors at play, it is difficult to be confident of the outcome. This should, though, not be used as a catch-all excuse.

In simple terms, if a valuer considers a property to be worth £100,000, but at the time of the valuation the comparable evidence suggests it is worth £50,000, then the valuer can be held liable. However, if the property *was* worth £100,000 at the date of valuation, but as a result of a downturn in the property market, it was subsequently worth £50,000, then the valuer would not be liable.

South Australia Asset Management Corporation v York Montague (1996). Contrary to earlier decisions, this case limited the valuer's liability to the overvaluation on the date of valuation. This has become known as the SAAMCo cap. As a consequence, a valuer cannot be held liable for losses sustained due to a downturn in the market *unless* he or she was expressly engaged to provide market and/or investment advice.

Basic market forces cannot be ignored, but these are not neatly packaged as with other commodities. Property is not homogeneous and records of sales experience throughout an area (beyond the limits of an individual firm) have, in the past, been poor. Now, with the widespread availability of sales data in the public domain, this process has become much easier. Developing a pattern of supply and demand even at the local level is, though, still complex and not a precise science. Anticipating future demand for a specific property has to be based on demand at that point in time and the number of similar products available that will be in direct competition. Consequently your clients take an inherent risk with any property transaction in respect of the intangibles.

The key features of the market analysis are summarised in Figure 8.6.

A further point that is of some significance in the residential market is the availability of finance, especially with a highly competitive mortgage market. There was a period in the late twentieth century when building societies had strict limitations on who should receive funds, based upon their own limited availability of finance. This put a brake on housing transactions. In more recent years, lenders have faced the imposition of strict affordability criteria on their customers, and this has had a similar impact, after many years of freely available and cheap mortgage finance that had fuelled rapid house price inflation.

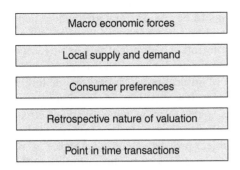

Figure 8.6 A summary of key considerations involved in a market analysis

8.4.5 *Calculations of price, worth and value*

How all the attributes influence price is the key to accurate appraisal and is the subject of much debate. A lack of detailed recording and assessment will mean that the surveyor's interpretation of value may be open to challenge. It can be argued that, for commercial property, the attributes will specifically relate to the needs of the investor. These are tangible and should lead to commercial gain. This underpins a definition of worth to the investor and there is no reason why this cannot be applied, to some degree, to the residential purchaser or lender.

The benefits perceived by purchasers of residential property are translated by the surveyor into the attributes and some of these are given in Figures 8.10 and 8.11 on p. 90. The attributes are identified on the basis of trends. A classic geographical example used in the northern hemisphere is that locations have developed into prime sites because of their position relative to the sun on south-facing slopes, as compared with north-facing ones. The perceived benefits for the purchaser may include a longer growing period for the garden, lower fuel bills and a more pleasant aspect. The number of benefits is great and varied and although one family's benefit (worth) may not be the same as another's, over time the benefits in a specific area will have grown to such an extent that a trend develops, and they become tangible, and this is translated as value.

This establishes the relative quality of the property within the marketplace and then a price can be applied, which reflects the benefits of owning or lending on that property in that specific area.

8.4.6 *Comparing transactions*

With any form of valuation there has to be a comparison, as value is relative. Whether you are valuing fine art or real property, the surveyor must have a benchmark. From this, good or bad points will be either added or deducted. There is a saying that suggests you should compare 'like with like' and so 'apples should be compared with apples, not pears'. However, here lies one problem with comparisons: although only one variety of apples should be compared with another, apples and pears are both fruits that are quite similar in some respects. The same can be said of property. A valuation should start by comparing properties of a similar type, e.g. terraced houses with terraced houses. But what happens when a detached house has to be valued in an area where there are mainly terraced houses and no other detached houses to compare it with? A form of comparison is still necessary to achieve a solution. There might not be any detached houses (apples) in the vicinity but are there some similar terraced houses (pears)? This is feasible, but requires a significant amount of subjective assessment by the individual surveyor. How the subjectivity can be reduced is described later in this chapter.

It is important that the surveyor bases a decision on what the marketplace is saying. A large sample of transactions makes for a large quantity of comparable data. Consequently the surveyor can relate the subject property against a number of others and rank it accordingly. A small number of transactions make this task more difficult.

Earlier in this chapter, we referenced the key attributes that we can use to compare one property with another. These factors, such as the type and style of the property, the number of bedrooms, aspect, proximity to facilities, etc. should be recorded either on the report or in the site notes, but this should give the surveyor a benchmark upon which to assess the value by comparison to other property. It will also act as a filter and, should the case prove to be complex, more detailed research can be undertaken. One methodical approach to this is to tabulate the features in a matrix, where you can easily compare one property with another. An example of such a matrix can be seen in Figure 8.7. The change from numbers 1–4 to letters A, B, C merely reflects that these may be less useful comparables possibly to put the subject property in context.

Address	Type	Beds	Recep	Baths	Gar	Value £	Size	£/sq mt	Date
Subject									
Comp 1									
Comp 2									
Comp 3									
Comp 4									
Comp A									
Comp B									
Comp C									

Comments

Tone of value

Figure 8.7 A simple comparable matrix that allows for tabulation and ranking of appropriate comparables

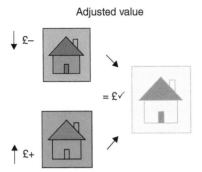

Adjusted value

= £✓

Figure 8.8 Graphic showing how the adjustment of attributes for the comparables properties can either be positive or negative. Green attribute means the comparable is better than the subject, so this produces a negative for the subject. Red attribute means the comparable has the better aspect and therefore this shows as a positive for the subject property.

By extending this matrix, the valuer can make adjustments to each of the property values to reflect the elements of the comparable property that are superior (adjustment downwards) or inferior (adjustment upwards). By adopting this approach, the valuer can come up with an 'adjusted value', within which range the subject property's valuation will lie, as shown in Figure 8.8.

By adopting this approach you will have clearly identified the basis on which you chose your comparable evidence and also the thought process behind how you arrived at your final figure. So, if the subject property has an average kitchen which is dated and in need of replacement, but the best comparable has a modern kitchen, then this is a positive that the comparable has over the subject, so the valuer would need to deduct the benefit of this modern kitchen from the best comparable to make it directly comparable to the subject. So, if the comparable is at £150,000 and the valuer considers the benefit to be £5,000, then the adjusted figure for the comparable would be £145,000. See Table 8.1 on p. 91 for further examples.

The experienced practitioner will take all the features for the subject property and consider whether, individually, they are better or worse than the best comparables (those most similar). For example, how close are the comparables from a location viewpoint? In the same road? In the same area, town or village? The greater the distance from the subject property, the more variables that are likely to arise. This will make the comparison more subjective and complex. The best comparisons would be property adjoining or neighbouring the subject one, reducing locational variances and allowing for a focus on more tangible items, such as number of bedrooms, size of accommodation and parking allocation. To use the apples analogy, comparison from one variety to another would be by appearance, texture, taste, keeping ability, price, etc. The principle is the same. However, these figures are not plucked out of the air, they should be based on market evidence. This only comes from analysis as it is currently not available on any database.

If, however, there are too many variables, then it is probably not the best comparable. Equally, the valuer needs to be selective on which attributes affect value. If one property is 3m² larger than another, does that actually make sufficient difference to adjust the value? A key attribute is a garage, but the relevance may vary. The fact that a house has space for a garage and therefore car parking is possibly the biggest determinant. This would be particularly true if a garage is an old prefabricated one, as opposed to a modern brick one or one that is integral to the main building, so constructed with cavity walls and therefore capable of an alternative use as extra habitable accommodation with some modification.

Let us look at how this may work out, assuming all these examples are properties on the same road, of similar size (95m²) and type, so have very few variables other than garage options, diagrammatically shown in Figure 8.9.

A typical house in this street is worth £250,000, with a space for a garage to the side of the main property. The valuer then needs to look at how to work out the impact of the three different scenarios given in Figure 8.9. The starting point is to look for comparable examples to see what market evidence there is. Ideally, this needs to be recent and in the same area, however, if that information is not available, then look further afield and look at percentage differences. How does a similar property with a prefab garage compare to one without or one with a brick attached garage? Is it 5 per cent, 10 per cent or how much? This is the starting point, building up these variances and getting a feel for how such attributes have affected people's buying decisions in the past.

> **Example 1** in Figure 8.9 will have a narrower plot and therefore less opportunity to develop, so this would most likely have the lowest value. For this example let us say the evidence proves 10 per cent less than the benchmark, so -£25,000. Value £225,000.
>
> **Example 2** in Figure 8.9 has space and a garage, which is possibly good for storage and is worth more than the property with garage space. However, is it the cost of building the garage that is the key to value? Possibly not, but what does the market evidence tell the valuer? It is likely that the difference will only be small, so 2.5 per cent more than the benchmark as the plot still has potential, i.e. +£6,250. Value £256,250.
>
> **Example 3** in Figure 8.9 has the same plot size as Example 2 but the garage is brick and attached, probably not cavity construction, but there are many examples of these garages being extended for utility rooms or kitchen extensions. Again the valuer needs to look for the market evidence and let us suggest in this case it adds 6 per cent, so £15,000. This is not necessarily a cost issue, but the benefit of what has been constructed to the buyer and more importantly, to the market. Therefore, we are looking at this example being priced at £265,000.

What the examples do show is a clear ranking as a result of the attributes created by the different garages and plot sizes.

These are only examples based on possible scenarios. There could be quite a few nuances that modify the figures, but hopefully it can show how the principle of adjustment can work, however, the key is building market evidence for such key attributes.

Where a property has an integral garage of the same construction as the house, then other considerations come into play. Not only is it likely to have a bigger plot, but there is the potential for more accommodation. On the basis of the base footprint of 95m², the benchmark property has a £/m² of £2,632/m². The integral garage plus additional accommodation above will be around 25–30m², but the finish of the garage will

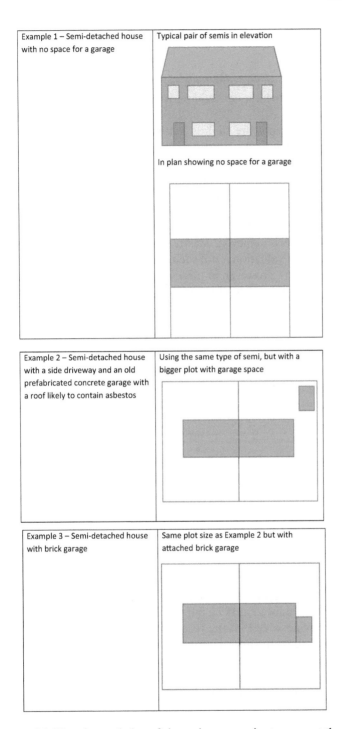

Figure 8.9 Elevation and plan of the various scenarios to support the comparable analysis

not be up to the standard of the habitable accommodation so that figure needs to be discounted, but it is better than the benchmark, so could be adding £40,000.

Once all the attributes have been determined from the selected comparables, then they can be ranked as shown in Table 8.1. Table 8.1 shows comparisons for a 2-bedroomed flat in the London area and the comparisons have been ranked on the basis of their similarity to the subject. As expected, the best comparison is one in the same street that is virtually identical.

The bathroom and kitchen fittings are slightly better in the subject so this would suggest that if the first comparable had better units, then it is likely to have sold for slightly more. So to make the adjustment for the value of those benefits an amount is added to the comparable price. Again, using market evidence built up from examining many sales transactions over many years, this will become intuitive. However a rationale always needs to be kept.

Identifying the key variables applied to each comparable, an adjustment is made, so that a range of adjusted values is achieved. Ideally, after adjustments, all the values should be the same or similar, but valuation is not such a precise science so do not worry if there are a few bigger variations. However, in this case the flats seemed similar, there were size differences, and some did not have garages. By making appropriate adjustment it is looking as though the property value should be in the region of £305,000. For 2-bedroom flats that are similar, then the valuer should be able to get a fairly precise valuation in a stable marketplace.

Ranking is quite a useful tool, because on many occasions the adjustments do not produce such a desirable result. especially where properties differ substantially. So in the same way as we used Figure 8.5, there needs to be an assessment of which properties are the best and which are the worst in value terms and then slot the subject into the value table. This then gives a good starting point for the parameters of value. The subject property is likely to be worth less than the better property but must be worth more than the property that is not as good.

We will develop these tools as we progress through the book, however, we have targeted the analysis at quite a simplistic level. As technology develops, using these principles with the extensive data now available will allow a much more sophisticated analysis to take place. The Automated Valuation Model is just one such technique, constrained by the limited attributes collected, but a combination of the eyes and ears of the surveyor and those algorithms creates a whole new world for valuation (see Chapter 10).

The Popular Housing Forum (a body comprising architects, planners and builders) produced a report entitled 'Kerb Appeal: The External Appearance and Site Layout of New Houses'. This report was produced in 1998, but the headings do not appear to have changed much. However, it is possible that the relationships between them may have moved slightly, so we have retained the information and added some specific commentary from the media gurus of the housing market.

This research identified some of the characteristics that appeal to purchasers and illustrates how a surveyor undertaking a valuation for a lender will not always reflect the same priorities as the purchaser (see Figure 8.10, p. 90). This graph identifies the reasons for preferring an existing property as compared with a new one. A surveyor would use size as a key attribute and the amenity of a garage for example. However, attributes such as character, attractiveness and that 'homely feeling' will be more difficult to quantify and maintain as a sustainable feature. The basic design of a property helps to establish

Table 8.1 Comparable sales analysis matrix

Comparable address	Property type sqm (if known)	£price/date; £/sm	Differences +/-	£ -/+	£Adjusted Price/Value
SUBJECT PROPERTY: 1 Ambridge Road	Flat 51 sqm Block of 4 Ground floor with garden area and garage 2 beds	£325,000 £6373/sqm			£373/sq £373/sq mo
23 Ambridge Road XX1 3BY RANKING = 1	Flat 52 sqm Block of 4 Virtually identical but needs some modernisation Has a garage	£299,950 £5768/sqm	- Needs upgrade kitchen - Needs new bathroom	+2k +5k	£306,950
64 Coombe Crescent, XX1 9BU RANKING = 2	Flat 51 sqm Block of 4 Very similar but needs more modernisation and has no garage space Slightly better street	£300,000 £5,882/sqm	- Needs new kitchen - Needs new bathroom - No garage - Location not as good	+5k +5k +15k -20k	£305,000
12 Tree Crescent, XX1 3RS RANKING = 3	Flat 60 sqm Slightly bigger	£310,000 £5,166/sqm	- Subject has nicer garden - No garage - Size adjustment 9 sqm x £5,166 = £47,000 but market evidence suggests £25,000	+5k +15k -25k	£305,000
14 Alice Road, XX1 3AS RANKING = 4	Flat 59 sqm 2-storey block Slightly larger but no garage and garden remote	£299,950 £5,083/sqm	- Garden detached from property - No garage - No private entrance - Size adjustment as per market evidence £25,000	+10k +15k +5k -25k	£304,950
24 Alice Road, XX1 3AS RANKING = 5	65 sqm 2-storey block No private entrance, no garage, but larger	£325,000 £5000/sqmt	- Nicer kitchen than subject - No private entrance - No garage - adjustment Market evidence suggests Less than straight calculation, say, £35,000	-5K +5K +15k -35k	£305,000

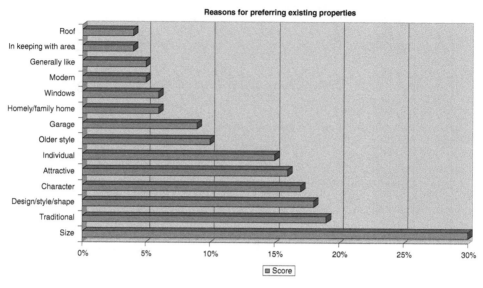

Figure 8.10 Graph showing the 'kerb appeal' of properties
Source: Popular Housing Forum (1998).

Fashionable 'in' features	Features 'out' of fashion
Broken roof lines	Boxy shapes
Eaves	Futuristic design
Bay windows	Space age design
Dormer windows	Unique
Varied use of materials	Ultra modern
Integral porches	Stuck-on features such as porches
Character properties	

Figure 8.11 Fashionable features in domestic house design
Source: Popular Housing Forum (1998).

the homely aspect, but it is quite often the choice and use of furnishings and decoration, together with general tidiness that actually determine the homely feel. This of course will change with a new occupier.

Fashion can affect the appeal of housing. The report by Kelly (1998) also listed 'the ins and outs of housing fashion'. This identifies an important factor that will change with time. Fashion in domestic design is quite likely to change a number of times within the lending term of the standard mortgage. Some of the key influences in 1998 are shown in Figure 8.11. Originally from the Popular Housing Group, apparently no longer in existence, but the information is still valid as a benchmark. The valuer should be asking: what are the current design features that add value? For example, it used to be a simple conservatory, but a recent development is the opening up of kitchen dining areas so that sliding patios doors open up to bring the garden into the living area, with lots of light and space having ultramodern features.

Research indicates that fashion changes over time and some research reported in *The Telegraph* in 2009 produced by Anna Tyzack, with support from Sarah Beeny, looks more at the tidiness and condition of the property, so:

- a nicely painted front door, even a 'wild' colour
- good quality door furniture
- good lighting around the door
- well-swept drives and paths, with no clutter
- numbering or naming houses in a good way
- clean windows, but positioning of window bars is important
- some greenery, if possible
- an impressive entrance gate (one for the larger properties)
- sprucing up a tired façade
- the neighbouring property can be an issue, especially if untidy.

The issues are shown in these photos: the bland grey-coloured bungalows for elderly residents in Figure 8.13 and the better affordable homes in Figure 8.12, that have the broken roof lines and stone effect walls that look much more appealing.

Phil Spencer, famous for *Location, Location, Location* and other similar TV programmes, commenting on the Zoopla site, gives the same sort of advice. However, all this is cosmetic, so is it that significant for valuers who need to see tangible results?

Figure 8.12 Showing speculative affordable housing made to look more attractive with broken roof lines and protruding gables housing the bedrooms

Figure 8.13 Standard grey-looking bungalows with no attractive features

This is why we have retained the earlier advice. You need to remember that a good or a bad look can easily be remedied or destroyed and so it is the core look that is important.

8.4.7 Comparing condition

Another area of particular significance in this book is the comparison of condition. This leads to even more subjectivity, as individuals have different views on the acceptability of cosmetic or minor defects. A newly decorated house will generally look more attractive than one that is well into the maintenance cycle and in need of painting. However, this is a cosmetic feature, as decoration does deteriorate with occupation and the normal ageing process. New decoration will carry a premium on the price (provided it is of a reasonable quality and colour) but whether it enhances value is debatable. Decorating is an ongoing cost and if it has been done recently, then the cost will be deferred for a further period. This will give benefit to the new owner and so add value. If the new owners fail to redecorate, then the added value will be lost, therefore decoration is not sustainable.

Where a property has more serious (material) defects, they may be referred back to the lender as conditional if the mortgage is to proceed. The repairs will usually be costed and reflected in the value. Table 8.2 (pp. 94–95) is an example of a comparable record of how condition affects the valuation process. The comments in each section provide the thought processes that can be adopted by a surveyor.

8.4.8 Comments on the comparable spreadsheet – a rationale

In Table 8.3, the key benchmark features have been used. Practices differ from region to region, office to office, but this example is probably the most common. It serves the purpose of evaluating the comparison method as a valuation technique. The headings are not 'tablets of stone' and can be extended, especially when linked to an integrated database capable of recording all the information on the valuation report or site notes. However, statistical evidence used in the House Price Index would indicate that there are only a limited number of variables that influence price in the majority of cases. There will always be circumstances when one of the key features does have an overriding impact. For example, the appearance of a property may be so awful that exceptional accommodation internally would be irrelevant, as potential viewers were put off before ever getting inside.

Table 8.3 (p. 96) has limited qualitative information. The norm is to simply list the information, though more sophisticated techniques have now been developed that rank the importance of the features. These databases effectively mimic what the surveyor does to value a property. It should be noted that the names of areas and streets have been changed for reasons of confidentiality and to avoid a regional bias.

Each of the categories is described in a little more detail:

Address The information shown merely gives the physical location. Implicit within that address are all those features that might affect value. For example, if coal mining affected the whole area, it would clearly need to be mentioned to a client. The surveyor would be so aware of the situation it would be a waste of time to record it every time. The exception might be where the mining was active or there was some particular hazard to that area, such as a known fault line. If there was a specific hazard, then this would be recorded in a separate comments field. Location can be ranked, as seen in the example of Celandine Nook. This area is considered less desirable than Longwood. More sophisticated techniques would score the area relative to other areas. Even within areas, it could be possible to score specific streets. In some cases this scoring might be applied to the actual plot. This only becomes relevant when there is a significant difference between plots. However, it is important to reflect that the plot is a significant part of the value of the property. In the south of the country, in certain areas, it can account for more value than the actual building.

In all the examples, reference has been made to the precise location and, in particular, the view of the dwellings. This is merely used as an example to differentiate one attribute from another. It is important as, in this case, all the properties are in a similar area. Therefore, this is fine-tuning the value, but in terms of lending risk, it is probably irrelevant. In terms of insurance risk, this feature may be more significant as exposed areas are likely to be more prone to storm damage or the influence of other climatic affects, such as frost. This is a good example of how the differing needs of clients can be significant. In this example, view is the key variable (see Figure 8.14) which shows the advantage to the eye and conceals the one of exposure, in another area it could be proximity to a railway or tube station, etc. Whatever the feature, the principles of comparison still apply.

Table 8.2 How condition affects value

Situation/classification	Movement	Damp defects	Timber	Roofs	Thermal insulation	Building services	Non-traditional dwellings
Decline for lending purposes	Structural repairs where the cost exceeds value – foundation failure in need of extensive underpinning and correction of levels within the dwelling to make habitable	Water damage has lead to secondary deterioration making property un-inhabitable – water damage from a fire or leak	Wood rotting fungi has lead to structural failure – dry rot extending to main structural elements and caused failure as with movement	Failure of roof structure to perform leading to structural failure – gale damage or failure of roof covering leading to water penetration as for damp, or failure of roof structure causing main walls to bow (excessive roof spread)	Cold bridging on building incapable of remedy, leading to condensation and mould growth that makes property a health risk – block of flats where problem inherent in the design, but cannot be resolved by one occupant	Failure of drains leading to structural failure of the main building – as for movement	A designated property where costs of repair to an acceptable standard exceed value – a less well known type for which there is not an economical repair scheme
Further investigation and impractical to provide a current value	Recent movement causing impairment to the use of the building – opening doors and windows restricted	Water damage has lead to secondary deterioration with the possibility of dry rot infestation – as for timber	Potential for dry rot or a failing of a timber element due to jointing, shrinkage, warping etc. – Timbers in contact with damp materials and mostly concealed	Failure of roof structure to perform adequately, a milder yet serious version of decline – roof spread where full impact of damage cannot be assessed as structural calculations necessary to determine extent of repair	Inappropriate use of cavity wall insulation or milder version of the decline scenario where a repair might be practical – use of cavity wall insulation in an exposed position making the outer skin too cold and risking frost damage to the bricks	As for movement, but caused by drain failure	A property where a repair has been undertaken, but it is not known whether it meets the standards specified by an approved scheme

Further investigation required, but quantifiable	*Recent movement of a less serious nature, where a diagnosis can be made from one inspection – crackingneeds toberepointed to prevent damp penetration*	*Evidence of damp penetration where the cause can be diagnosed and it does not extend to concealed areas – rising damp to main walls where the floor timbers can be seen in the cellar*	*Defect identifiable – windowframes affected by wet rot and needing replacement as the only feasible method of repair*	*Defect identifiable and actually failing lead to ingress of water – flat roof in need of re-covering and underside of roof visible so extent of damage apparent*	*Incorrectly installed insulation leading to cold bridging and serious condensation – situations around window or door openings in contact with timber elements*	*Evidence that the services pose a risk to health and safety – obvious defects or old installations for electric wiring*	*A non-traditional dwelling not designated and repairs are typical – spallingofrender on a Laing-Easiform exposing the reinforcing*
Defects form part of a programme of routine maintenance	*Old movement that will need regular repointing – thermal movement under a window, where the cracking does not extend below the dpc*	*As above, but the extent of damage is visible and the cost of works minor – defective flashing to a chimney stack and the underside is to a roof void*	*Examples of wet rot on secondary features of the structure – barge boarding*	*Evidence of failure allowing minor water penetration that does not impair habitation – slipped tile*	*Inadequate levels of insulation – insufficient loft insulation*	*Minor defects that do not pose an immediate threat to health and safety – one old plug socket with surface wiring*	*Repairs typical of a traditional dwelling*

Table 8.3 Example of a comparable spreadsheet

Address	Type	Bedrooms	Age	Size m²	Tenure	Condition	Garage	Price	Value	Date	Extras	Analysis value m²
2, Acacia Ave, Longwood, Anytown. Good views; faces the south, back garden in shade, but a corner plot, lacks privacy	SDH / Typical for the area!	3 / 2 large 1 small	45 / typical	90 / average	F/H / usual	Good / average for this area	Space / usually a garage	75000	75000	03/09/97 / On the market for 3 mths usual for this area	CH, double glazing / Usual	833.33
53, Acacia Ave, Longwood, Anytown. On the other side of the street, good views from back garden, better position than 2	Same	Same	46 / Similar	90 / Same	F/H / Same	Fair / Cosmetically worse than 2	Garage / Pre-fab, better than 2	80000ᵃ / Same	77500	10/10/97 / Little movement in the market place in this period	CH, double glazing / Same	861.11
24, Lilac Road, Longwood, Anytown. A street parallel to Acacia Ave, not as good views, same access to facilities	SDH / Same	4 / Extended property	40 / Younger but not significantly same construction	105 / Larger	F/H / Same No 53	Good / Same as No 2 better than others	Garage / Brick built better than others	92500ᵇ / Same	90000	15/09/97 / Same	CH, double glazing / Same	857.14
18, Ash Lane, Celandine Nook, Anytown. An area adjacent to Longwood, but higher up the valley side, more exposed and more urban views, similar access to facilities	SDH / Same	3 / Same as Acacia	35 / Same comment as Lilac	92 / Similar	F/H / Same	Good / Similar	Garage / Same as Lilac	70000 / Same	70000 / Lower price reflects location	23/09/97 / Same	CH, double glazing / Same	760.87

a Higher price reflects better position and garage, but it could have been higher if condition better. b Higher price reflects larger accommodation, better garage, but poorer position. General comments on Longwood: good area, formerly wooded, some evidence of longstanding movement, local shops and a good school nearby. Access to a motorway, good views of a golf course and valley and to nearby hills, exposed in winter

Figure 8.14 The view from this property is both an asset and a liability. The scenic outlook will enhance value but the exposure to wind and rain may affect the durability of building components.

The aerial view (Figure 8.15) shows the variables of property location. In the lower part of the photograph there is a development of good quality stone dwellings adjacent to a dual carriageway road. To the upper part of the photograph there is a supermarket and petrol filling station. Beyond this is a council housing estate. The property fronting the main road is of an older type to those on the private estate and the plots are generally smaller. The relative value per square metre of these plots will be lower because of the proximity to the main road and the (less than ideal) view of the supermarket. However, the latter point has to be balanced against the convenience factor.

A further point relating to location is occupational mix. It is sensitive, but cannot be ignored. In this case there are two forms of housing tenure in close proximity. Those in private occupation and those rented from the local authority. It is not uncommon to find property of differing tenures that are close to each other. In some cases, developers are required to allocate part of a new site for social housing. This can result in families of different socio-economic groups living in the same area. It can also provide accommodation for those not choosing to buy. Whatever the reason, this will be interpreted by the marketplace. This could lead to a reduction in the capital value of the owner-occupied property and result in a higher rental value for those that are let. Each case needs to be looked at on its merits; the surveyor needs to be aware of the potential impact and allow for this in the analysis.

Figure 8.15 The aerial view of this small neighbourhood area shows how location can affect houses differently even though they are relatively close to each other

Type   In all cases on the comparable sheet, the type is the same. Implicit in this is that building costs will have been similar. It would be important to distinguish if the design differed markedly. For example, if one dwelling had a flat roof while another was thatched, there might well be an impact on value. Again this aspect may be more significant to the insurance risk than the lending risk, although appearance would be a key feature in assessing the value.

Bedrooms   The number of bedrooms gives an indication of the size of the accommodation and its functionality. (Americans would record the number of fireplaces, as this indicates reception rooms, which is of more significance to the customers in that country.) Within the range of 1–5 bedrooms it should be possible to relate this to the size of accommodation (but there are always exceptions!). Beyond this, any unique or special characteristics may require a more subjective approach. As a rule, a 3-bedroomed house will have a wider market than a 1-bedroom because there are usually more potential purchasers. The 3-bedroomed property is more adaptable and better suited to the average family unit. However, certain locations can favour a certain size of accommodation, a student area, for example, or one near a hospital, that may consist of traditional Victorian housing. It is quite likely that a proportion will have been converted to bed-sit or flatted accommodation rather than being occupied as family units as originally intended. The valuation process is a complex matrix of matching variables.

Age   This can give an insight into the condition and to some extent the design of the property. For example, a property built in the nineteenth century will not meet the same standard of construction as one built recently. Therefore it would be unusual for it to have the same level of energy efficiency. If it did, this would be considered an advantage, especially when compared with properties of a similar age that would almost certainly be energy-inefficient.

Square metreage   This indicates the size of the property, but the surveyor would need to give thought to how that space has been used. For example, a room of 5 m × 5 m with appropriately positioned doors and windows, giving adequate access and natural lighting, has more potential than one of 7.5 m × 3.5 m. Such a room would be long and narrow, it would be difficult to fit the furniture in to a traditional layout. Rooms can be too big and feel cold and draughty. A hallway no larger than a passage between doors would give a better appearance, where the stairs lead up to the first floor, as this also gives an impression of more size.

Tenure   There are numerous legal aspects that can influence the value; some very significantly (see Chapter 5). Tenure is a significant one worth recording as the differentiation between freehold and leasehold is the most important. Other aspects are so wide-ranging and exceptional that they would be recorded and interpreted individually.

Planning   A legal consideration. For an established area, the Planning Authority defines the policy and therefore close comparables will not vary. The main example is likely to relate to any alterations to the property and whether they comply with current regulations. Breach of planning regulations is likely to be exceptional but significant.

Condition   These are all-encompassing one-word comments that have limited value to anyone other than the individual who made them. The main use is as a memory jogger. It is important to set a common scale so that a benchmark is set.

- 'Excellent', i.e. comparable to a new house.
- 'Good', would suggest that there were no repairs other than would be encountered within normal maintenance and that the general appearance was tidy.
- 'Fair' or some other equivalent word meaning less than average, would need some works. Some of these may be made a condition of the mortgage.
- 'Poor' would indicate a number of repairs required.

Construction   Different types of construction will be discussed in Volume 2 of the book, but we need to consider the impact upon value. As already specified, the basic requirement is to compare like with like, so if the property is system-built from shortly after the Second World War, then it would not normally be ideal to compare it with a traditionally built property of similar age and size. However, it may be possible to establish a tone of value for an estate of similar constructed system-built houses with an estate of traditionally built houses and so the differences can be established and therefore accounted for.

Beware those properties legally defined as defective under the Housing Defects Act 1984. For these to be accepted, there are defined schemes of repair available that make them effectively traditional, but there are many copy-cat schemes that do not necessarily meet all the criteria of the official schemes. If not repaired, or repaired in the correct way, then the value will be significantly affected as most could be un-mortgageable.

Increasingly, new methods of construction are entering the market. Often referred to as MMC (Modern Methods of Construction), many of these are untested over time. The Build Offsite Property Assurance Scheme (BOPAS) and NHBC Accepts are two recognised certification bodies' schemes that test the durability of these new construction types and consider if they have the longevity to be considered as mortgageable. It can be difficult to identify MMC properties without thorough investigation as, in many cases, they look the equivalent of a traditional house. Indeed, some estates may have a combination of different construction types that all look outwardly standard in construction. Where a property has a traditional look, the market may not differentiate on price, but it is important that the valuer recognises the construction type and makes the client aware of any actual or potential issues that may influence future maintenance and durability.

Energy efficiency   Most modern constructions are designed to comply with ever stricter insulation requirements and are now more energy-efficient. This will translate into lower energy bills and running costs which should equate to a value-added incentive. However, at the time or writing there is no clear evidence that this is the case, or how much this may be.

On second-hand houses, the Energy Performance Certificate (EPC) makes recommendations for energy improvements and costs their potential benefits. Apart from the least expensive items such as loft insulation, energy-efficient light bulbs and cavity wall insulation, the pay-back period is such that it is hard to rationalise any significant influence on value. However, with increasing emphasis on the need to achieve a carbon-neutral status, then measures may come in that have some influence. If a certain type of heating boiler is banned, then there is likely to be a negative impact on the value of a property with such a boiler to allow for the cost of replacement. In this respect, watch the impact on car prices as they seem to lead the way on energy monitoring.

Garage   This covers number of garages and parking spaces. In some areas the ability to park off the street will carry additional value, because on-street parking is the norm. In the examples quoted, off-street parking was standard for the area.

Price   The price quoted in the marketplace at the time of valuation. The surveyor may wish to ask a number of questions about this:
- How long has the property been on the market?
- Has the price changed over the period?
- How many people have looked around the property?

The answers may not always be forthcoming, but if they are, then they give an indication of how the customers feel about this property. Clearly, if the market is very good, then the customers may not have a lot of choice, dependent upon the degree of their need. If the market is very poor, then it is likely that only the best at a price will sell. Again, these are extremes, and the normal situation is more likely.

Value   This is the surveyor's interpretation of all the features put together when compared with similar properties. The reason for the valuation may dictate how this value is recorded (see the definition of value and worth earlier in the chapter). The courts have indicated that there are varying degrees of tolerance that are acceptable. Care must be taken in interpreting such decisions, as they will all have been based upon the specific facts of the case. To use the 10 per cent example could be critical on a 95 per cent mortgage, especially if comparable evidence exists for a neighbouring property, as this could well be a near-perfect match. In such cases, 10 per cent would be excessive. Remember it is for the courts to exercise tolerance based on the evidence presented. It is for the valuer to provide what is considered an accurate valuation at the time.

Use of comparables

Blemain Finance Ltd v e.Surv Ltd (2012)

The subject property was a top end residential property, first purchased in 2004 for £1.92m. It was valued by e.surv in 2007 for £3.4m. It was found that £2.7m was the correct valuation and that inappropriate use of comparables had been made. This inappropriate use took various forms including that defendant did not know the size of all comparables, incorrectly stated the sale price of one of the comparables and stated they were in a worse location than the subject property when, in fact, they were found to be in a better location.

Redstone Mortgages v Countrywide Surveyors (2011)

This case involved the re-mortgage valuation of an end-of-terrace house located on a service road parallel to an exceptionally busy A road. It was a matter of common sense that the primary comparables should be properties on the same road. Of the 27 comparable properties referred to, just five fulfilled this criterion. In addition, the surveyor had referred to his firm's own database when selecting comparables and had, in fact, himself valued a number of the comparables. The judge noted that far more care should have been taken to find and use comparables from other sources.

In short, the courts highlight that there must be a clear and valid justification for using any comparable.

The RICS have given guidance (*Valuation of Individual New-Build Homes*, RICS, 2019a) on the procedure for valuing new property, which relates to the number of comparable sales actually on the site and comparison to other sites and second-hand sales in the vicinity. The surveyor has to determine, on the circumstances of each case, whether they have effectively proved the figure reasonable and that it provides a suitable risk for the lender. This may mean that there is a tolerance between the comparable evidence and the price paid in the marketplace. This reflects the market at that point in time. For example. the difference between the best comparable from a month previously may show a 5 per cent increase in value. If the market is fairly active, this change may be acceptable and to ignore it would be incorrect. Provided the surveyor records the reasoning behind the decision and this is justified, then the tolerance is acceptable. The reasoning behind the process of fine tuning the valuation is complex. Any differences could be due to a variety of factors ranging from the time of year (the market is usually quieter around Christmas) to the demand for particular properties in popular areas, or the impact of government initiatives, such as Help to Buy, or obligations placed on the development, such as section 106 requirements (Town and Country Planning Act 1990).

Date   The date the valuation was completed should also be the date of the inspection. This is important to establish a reference point. It is unlikely that anything would change between the day inspected and, say, a valuation worked out a day later. If the house did burn down or was damaged in some other way and the valuation was dated after that day, then there could be an issue.

Extras or comments   A catch-all section to cover any additional or overriding factors that will influence the value. Central heating and double-glazing are recorded here because, without them, the value would be significantly affected. In respect of central heating, this is probably by as much as, or more than, the cost of the installation.

Analysis   To illustrate the comparable method, a simple analysis has been included in Table 8.3, but there are many variations. This example highlights some of the difficulties and the subjective nature of the analysis. All the figures vary. That is because the figure has to reflect not just the building, but also the immediate and extended environment, condition and all the other features. Surveyors in different parts of the country may argue with this but it must be remembered that all these figures are relative and therefore the use of percentages may be more appropriate. The variances can only be arrived at by analysing previous examples and a key feature is to link that analysis to price and to keep doing that. Failure to monitor the market would lead to a self-perpetuating situation whereby the surveyor would have a stable or even declining set of figures.

In short, the valuation of a property comes down to a step-by-step process and a comparison by numbers:

1. Define your neighbourhood where there are common characteristics.
2. Identify whether there are key facilities in that neighbourhood that make it more or less attractive/popular to the buyer.
3. Identify the tone of value in that neighbourhood.
4. Identify properties that have sold recently that ideally match the subject, i.e. size, style, type and key attributes.
5. Analyse. If not an ideal match, then make adjustments for those attributes that do not match.
6. Produce the rationale to show why the comparables support your valuation

8.4.9 A review of factors that affect value

This section will review, by reference to recent research and practical experience, how clients perceive the benefits that accrue from expenditure on their property. In trying to understand the impact of condition upon value, it is important to set this within the context of other issues that affect value.

Interpreting the value of the benefits to an individual property tends to come from practical application. Very little has actually been written about the process in the UK, probably because the variety of benefits is so great that they are not well recorded. If they are, they are jealously guarded because these factors form the basis of the surveyor's art and business. Where property is similar (i.e. a row of terraced houses), then there may be few attributes that differentiate one property from another. The sunny side of the street may be known to be of significance to locals. Where there are only small back yards, then orientation to the sun is very important. It is probably of less worth to the occupier who has a larger garden where the shadows do not extend over the whole plot. Conversely, local knowledge about a former resident may also be sufficient to give the house a stigma that will make it unsaleable. The key learning point here is that benefits are of differing value to differing individuals and this may be related to location or other attributes.

8.4.10 Home improvements and value

In May 1997, 1,200 questionnaires were sent to estate agents throughout the UK (*Which?*, 1997) asking whether 20 popular home improvements would influence the price and saleability of a home. The findings are shown in Table 8.4. What tables such as this do, is to continue to give a benchmark so that, as changes happen, then consideration can be given as to why and how this has affected the valuation process.

A literal interpretation of these findings is inadvisable. This is because the appreciation in value of a property is more likely to be associated with an improvement in the general property market rather than as a direct result of a single home improvement. House prices have on average more than doubled within the last 30 years (even allowing for the recession in the late 1980s to early 1990s). Consequently £1,000 spent on central heating back in 1983, on a house worth £50,000 would have been lost within the overall appreciation of value. The average value could now more than £100,000. This inflationary action has been active since the post-war building boom. For example, a 4-bedroomed detached house in South Yorkshire newly built in the mid-1960s was originally bought for £4,500. In 1997, it was sold for £105,000 (2,233 per cent increase!); the same property may well sell in 2021 for £400,000. Future predictions are for a better-managed economy and for a more stable level of inflation. Consequently, it would be unwise to rely too heavily on the market to overcome errors in judgement on home improvements. The following guidance may be helpful.

A home improvement should:

- **be of benefit to the actual owner and should give worth to that individual**. If the improvement can be replicated cheaply, then it is unlikely to add value. An example would be if, instead of building an extension to a house, the same amount of accommodation could be acquired by buying another house more cheaply. Alternatively if a neighbouring property has had its windows replaced by PVCu double glazed

Table 8.4 Results of whether home improvements affect the value of property

How home improvements can affect value a lot	How home improvements can make a house easier to sell
Will increase the value a lot	**Will make selling much easier**
1. Two-storey extension	1. Central heating
2. Central heating	2. New kitchen
3. Garage	3. Garage
4. New kitchen	4. Off-road parking
Will increase the vaue a little	5. Double glazing on a modern house
5. Conservatory	6. New white bathroom suite
6. Off-road parking	7. En-suite bathroom
7. En-suite bathroom	**Will make selling a little easier**
8. Loft conversion	8. Paint exterior woodwork
9. Double glazing on a modern house	9. Landscape gardens
10. New white bathroom suite	10. Conservatory
11. Landscaped garden	11. Two-storey extension
12. Original fireplaces	12. Original fireplaces
Will make no difference	13. Redecorate in neutral shades
13. Replace sash windows with PVCu ones	14. Loft conversions
14. Redecoration in neutral shades	15. Hanging baskets and window boxes
15. Swimming pool	**Will make no difference**
16. Paint exterior woodwork	16. Roof insulation
17. Roof insulation	17. Replace sash windows with PVCu ones
18. Redecorate in distinctive shades	18. Redecorate in distinctive shades
19. Hanging baskets and window boxes	19. Knock through rooms
20. Knock through lounges	20. Swimming pools

Source: Which? (1997) Published by the Consumers Association. www.which.co.uk.

windows and those of the subject property are single glazed and in poor repair, then the value of that property is likely to be depreciated by a sum to reflect the cost and inconvenience of replacing the old windows. Situations are rarely as clear-cut as this, but this gives an indication of the principle.

- **be of good quality and the appearance should be in keeping with the design of the actual property and of those around it**. The classic flat-roofed two-storey extension on a 1930s semi-detached house may give benefit to the owner but its appearance is not up to modern-day expectations. Therefore its added value will be minimised. The addition of a pitched roof will be expensive but will certainly enhance the appearance. It could also reduce maintenance costs and therefore will add some element of value. Whether this will equate to the money spent on the improvement will have to be judged in the individual circumstances. Another aspect would be where the extension greatly increases the size of the accommodation to the extent that it is not characteristic of the type and socio-economic grouping of the person who might want to live in that area. Then this will not add value, merely benefit the individual owner. The same applies to the detached property positioned in an area of terraced houses.
- **should provide additional functionality**. One example given by the Consumers Association (*Which?*, 1997) shows a conservatory on a reasonable quality bungalow owned by an elderly couple. They considered the main benefits were 'a pleasant place to have their dinner and entertain friends'. The value of the property was put

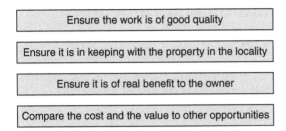

Figure 8.16 Summary of guidance when assessing the relative value of home improvements.

in the region of £145,000. This represented an increase in value (partly due to infla-tion) of £15,000 which covered the cost of the investment (11.5 per cent). If this had been a £30,000 terraced house, expenditure of £10,000 (33 per cent) or more on a conservatory forms a high percentage of the base value. It is unlikely to give that degree of added value, especially if the durability of the conservatory is considered. A brick-built dwelling has a minimum life expectancy of 60 years, whereas sec-ondary features such as softwood window frames have, probably, a life expectancy of 10 years. A typical conservatory falls somewhere between the two. Consequently the terraced house owner has to look carefully at what level of functionality can be achieved by the type of improvement.

A summary of guidance to the surveyor or owner is shown in Figure 8.16.

9 Reporting on condition and value for a residential mortgage

9.1 Reporting overview

The main purpose of this chapter is to concentrate on the influence that condition has on value. However, it is also the chapter where we will reference the production of the general remarks section for lending purposes. This is complicated by the fact that lenders differ in the information they wish to collect. Usually, they are a combination of simple tick box sections which highlight the core information that the lender needs and may coincide with the underwriting programme run by the lender's computer system. It is critical that these boxes are used correctly, because they may well determine if that report form will even be looked at by a human being. All the valuer's hard work could be lost by just one small typing error. Therefore, we cannot say much about the multitude of different ways a lender collects information.

9.1.1 General remarks in reports

We do need to consider the area often known as General Remarks. Even then, some lenders do not have such a section, and some insist on standard phrases being used. In the latter case it is important to ensure that the phrase fits the situation, and that the situation is not adapted to fit the phrase.

Where there is no opportunity to highlight something to the lender that the valuer considers would have a material impact upon the valuation, and the valuer also wants to exercise the duty of care to the borrower, then the valuer should make every effort to inform the lender in writing of the situation. This may be something like the need to check that a recent extension has the appropriate approvals, because without them there could be an enforcement action and the need to remove the extension, which would obviously materially impact upon the value.

> The significance of following professional guidelines rather than system-driven requests from a client were highlighted in the case of *Webb Resolutions Ltd v e.surv Limited* (2012) .
>
> As part of this case, consideration was given to the terms of the agreement between the lender and the surveying practice, but also reference was made to the RICS Red Book relevant at the time the valuation was undertaken. The defence barrister had expressed concern that the valuer was unable to give the appropriate information to the lender as there was no provision for general remarks within the report form, which was in accordance with the lender agreement.

DOI: 10.1201/9780367816988-9

The case quoted the RICS *Appraisal and Valuation Standards* (which were incorporated into the Supplier Agreement by clause 5.6) which include, at appendix 5.1, a document entitled 'The Minimum Contents of Valuation Reports'. This includes the following:

> The valuer must make it clear if the valuation is being carried out without the information normally available when carrying out a valuation. The valuer must indicate in the Report if (where practicable) verification is needed of any information or Assumptions on which the valuation is based, or if information considered material has not been provided.
>
> If any such information or Assumption is material to the amount of the valuation, the valuers must make clear that this valuation should not be relied on, pending verification …

As a consequence, the judgment included the following:

> An issue arose, in particular concerning the valuation of 1207 Masshouse, the property purchased by Mr Ali, as to what the e.surv valuer should do if his obligation to take reasonable care (and/or his specific obligation to comply with the RICS guidance notes), required him to do or say something which could not be accommodated by the tick boxes provided. I am in no doubt at all as to the general answer to that: if the valuer could not discharge his contractual or tortious obligations to GMAC without adding words somewhere on the form, or producing a covering letter to address the particular issue, then he had to take that course, regardless of the warning in the GMAC guidance notes; anything else would be a plain failure to comply with those obligations.

Therefore, we would advise in generic terms that reports need to cover the following:

- The starting point for completing any General Remarks is to follow the guidance as set by the lender, provided the circumstances of the case support that as the appropriate professional and technical approach.
- Ensure that the information provided to the client is relevant and not duplicating information already provided in the tick boxes.
- Set the scene for the underwriter and the applicant to the mortgage so the property is in context.
- Identify essential items of repair or other matters where action is required that could influence the lending, saleability and therefore the valuation. These are matters that the valuer considers need to be imposed as a condition of the valuation. It is for the lender to determine whether they will be a condition of the lending. Some lenders allow two valuations: one for the property in its current condition, with the current circumstances, and another one when works have been completed or assuming other actions resolved.
- Consideration needs to be given to the applicant to the mortgage as the valuer has a duty of care there as well. The valuation is a limited inspection so not all defects or

issues will be discovered but there might be something such as a health and safety issue that should be mentioned. Electrical and gas safety checks may, on the face of it, be a formality, but as the valuer is not a specialist and has not undertaken any tests, then there may be something that is concealed and only by having the test can this be discovered, which may save someone's life. The cost of the test and the repair may be small in relative terms to the price of the property, but life-saving. This is a judgement decision on the day of the valuation.

The next section looks at condition so we will try to highlight those areas where a comment in the report would be useful.

9.1.2 Reporting on condition

First, a distinction can be drawn between home improvements and condition. In some cases they can be the same, as with the example of replacement windows. But in one important respect they are different. The works to retain the condition are essential if the value of the property is to be maintained. Provided they are done correctly (i.e. sustainably), then they will be of benefit to the owner. The owner may not agree with this, especially if repairs were unexpected and the money spent on them was allocated for other matters which gave a more tangible return, e.g. a family holiday or new car.

The effect of repair works on value can be as unpredictable as home improvements. One thing is for certain – a property allowed to fall into disrepair will lose value compared with a better maintained one. It is this latter point that is important and relates back to the definition of 'value' from the dictionary – a fair equivalent, the degree of this quality (see Section 2.2.1). In undertaking the appraisal of a residential dwelling, there are three levels of condition that need evaluation:

1. the defect that is so serious it will cause, or has caused, major structural failure (e.g. foundation failure);
2. the defect that is serious and could, if left unattended, lead to a partial structural failure. An example would be defective roof covering that is allowing water penetration. That could lead to rot occurring in structural timbers. In other words those defects that render the dwelling not 'wind- and water-tight'.
3. the defect that is not serious, mainly cosmetic, and would only affect appearance if left unattended.

Example (1) is so serious that the value of the dwelling would be materially impacted if left unattended, resulting in a total loss with demolition being the only option. In such cases, it is only worthwhile expending money on the property if the cost of the works does not exceed its completed and fully repaired value, excluding the land.

In respect of (2), the same criteria apply. In this case it is more likely that the nature of these repairs is such that they currently affect individual elements of the property. Therefore, the value of the property will be eroded by the cost and inconvenience of undertaking the repair relative to better repaired ones. If all the properties in the locality have the same defect, then the comparison on value has to be spread to other localities to find comparables that are in better repair. In market situations where there is pressure on supply, then this cost equation could be reduced, as individuals will have differing views on how to undertake the repair, or how to pay for it. Ultimately, the ability to pay influences value.

In the final area, (3), the costs of works may not be fully reflected in the price or value as they may be indicative of on-going maintenance and therefore inherent in the pricing of that type and age of property.

The impact of condition can also be a complex matrix of two or more attributes, which affect value. For example, condensation, which has a perceptible impact upon the occupier, can be a result of a combination of factors including location (colder region), the quality of finish within the dwelling (level of insulation in the dwelling), the type of property (non-traditional type) and the use and layout of the accommodation (high moisture production).

How all of this should be reported depends upon the needs of the particular client. The lender will want to know of high-risk situations that could influence the security of the loan. They will want these elements reduced, if at all possible, for example, the damp staining to a rafter, indicative of a leak to part of the roof that can only get worse and could ultimately lead to dry or wet rot. The potential for structural failure needs to be assessed and reported accordingly. The occupier may choose to review a report in a different light, dependent upon the perception of benefits that will accrue from the property. If the leak is not actually affecting the occupier's use of the property (i.e. the water is not dripping onto the bed), then they may wish to defer an expenditure in preference to another which will produce a more immediate and perceivable benefit, such as redecoration.

The folly of this decision is apparent to those who view the property in a more commercial way. Consequently it is important in all cases to consider the needs of your client and, where practical, ensure that advice is given in the right context and prioritised if appropriate. The following section gives an example of categorising the condition of a property that is being considered for a loan.

9.1.3 Distinction between lender and purchaser

Figure 8.8 in Chapter 8 shows the assessment of attributes relative to value. In this example, condition is not a major factor, but it is important to note that certain types of defect can or will have a material impact upon the value and saleability of the property and thus the decision to lend money. Situations where the surveyor is unable to give a value are rare, but they do occur. Special attention should be given to avoiding the need for unnecessary requests for specialist reports when valuing for a lender. Where the surveyor can make an accurate diagnosis on site and the full extent of the repair can be identified, then the impact upon the value should be determined.

Situations will always vary, and care should be taken in reporting defects. Whilst individual lenders may have specific instructions, as a rule, they are not interested in defects that are not material or will not cause the property to deteriorate, expecting the valuer to reflect general condition in the reported valuation instead. Figure 9.1 shows properties in one street that appear to be in a similar condition and therefore age and general condition will be reflected in the price/value. Where your client is a member of the public, take care not to report all defects and stray into completing a mini-survey, as this will create expectations that cannot be met and may lead to complaints.

Figure 9.1 Properties in one street in a similar condition

The following is a summary of the definition for the options used in reflecting condition in a mortgage valuation:

- **Decline for lending purposes**: In these cases the defects are so serious that they question the future stability or durability of the dwelling. There may be occasions where only a comprehensive refurbishment programme could provide a solution.
- **Further investigation and impractical to provide a current value**: The extent of the defects is quite serious and cannot be fully diagnosed, so it is not possible to quantify the damage and the cost of the works. The surveyor therefore needs more information to rank the property relative to its comparables. Figure 9.2 shows a tree-lined road. The trees are of some size and may influence the properties, especially if on a shrinkable clay soil. Where there was some evidence of movement to the property, then this is an example of a case where further investigation would be sensible. The vendor would be well advised to obtain independent reports before sale.
- **Further investigation required but quantifiable**: The surveyor can diagnose the extent of the defect and therefore quantify the likely cost so the needs of the lender can be satisfied. The purchaser will need to be kept fully aware of the implications of the costs of the necessary work.
- **Defects form part of a programme of routine maintenance**: This helps complete the picture of the type and general condition of the property, probably of limited significance to the majority of lenders but helpful to the purchaser and to the future saleability of the dwelling. May indicate areas that could give rise to an insurance claim in the future.

Figure 9.2 The proximity of trees to the properties should be of clear concern to any surveyor and a justifiable reason for recommending further investigations, where they are of a type and size that may cause damage and risk the household insurance.

In respect of the lender, the key areas of concern can be categorised as follows:

- dampness
- movement
- timber failure
- services
- health and safety
- cost and its impact upon value.

The significance of any defect is a reflection of whether it impairs the quality of life of the occupants. Some are related to the same cause, so they form a matrix that will combine to produce a plan of action for the surveyor. Questions must be asked as to whether the defects pose a risk to the health and safety of the occupants. If so, is this imminent or an acceptable future risk and therefore what is the impact upon value? How much will it cost to repair? The significance of this latter point will be with reference to how the property ranks against other property within the same area. The defect may be common to all property and be subject to an acceptable measure supported by qualified contractors. There are often damp problems in a 1900s terraced house and there are plenty of contractors capable of undertaking the work and the costs are well known, so there is a quantifiable impact upon value.

The purchaser needs different information to the lender. The lender is concerned that the loan can be repaid, and the property has some influence on this, but it is mainly the purchaser who provides the basis for the security. However, the purchaser has a need

Figure 9.3 No, this photograph has been printed straight! The front fence posts are vertical, it is the house that is leaning!

from the property as a home that may be impaired if the defects affect their occupancy and enjoyment of a house. It is important to draw a distinction between those who view the property in a more commercial way and those who view it as a home, consequently, it is important in all cases to consider the needs of your client and, where practical, to ensure that advice is given in the right context and prioritised. Figure 9.3 shows a property leaning at a considerable angle, built on sand; it may well provide reasonable accommodation for the occupier and they may have taken internal measures to ensure everything is on the level. However, saleability is restricted, and the commercial view must be taken if a loan is required. Would a prospective buyer prefer this house or one genuinely on the level? What reduction in price would they expect to compensate them for the movement?

Achieving a balance between genuinely persuading the client against purchasing a property that is not suitable for them, and understanding the extent to which a client may be able to deal with defects, is an art in itself. The key is that even a dwelling in the worst condition can be rectified by the right client. Therefore understanding your client is the most significant issue.

10 Alternative valuation methods and approaches

10.1 Overview

This book is primarily about the valuation of residential property using the comparable approach. However, it would be incomplete without dealing with the alternative approaches with which a valuer should be familiar. In this chapter, we touch lightly on the residual and the investment approach, which are of use in particular circumstances and also as a cross-check for some valuations. Practitioners who wish to understand more about these approaches should consult specialist textbooks or attend appropriate training.

This is an area where the sections of the Red Book become very important in respect of expertise.

Practice Statement 2 (PS2) 'Ethics, competency, objectivity and disclosures' states:

> As it is fundamental to the integrity of the valuation process, all members practising as valuers must have the appropriate experience, skill and judgment for the task in question and must always act in a professional and ethical manner, free from any undue influence, bias or conflict of interest.

So, what exactly does this mean? It states in PS2 1.4 that 'opinions of value prepared by a member having the appropriate technical skills, experience and knowledge of the subject of valuation, the market and the purpose of the valuation'.

It goes on to say that just because a valuer has a qualification, it does not imply that the valuer has the practical experience in a particular sector of the market. For the type of valuations we are now going to discuss, having the appropriate experience is vital to understanding how to apply the methodology. In some cases this may be achieved by working with another surveyor who does have specialist knowledge. We will therefore highlight some of the key issues that need to be addressed when looking at properties where a different methodology may be required

10.2 The Residual approach

The Residual approach to valuation is for situations where development is involved and direct comparables are not available. In all cases, using direct comparison is preferable to any adjusted valuation, but sometimes it is simply not practical to get comparables, so logic and common sense can be applied.

DOI: 10.1201/9780367816988-10

For example, the valuer is required to calculate the value of the land that is to be developed for a single plot. Whilst it may be relatively easy to find evidence to support the value of the completed property, comparables for the plot are much more difficult to find. Usually a lender needs to know what the starting point is, i.e. the value of the land, as they may need to finance the land or to use the value of the land as collateral for other elements of the development.

If that land is agricultural, then it may be possible to find comparable evidence for other pieces of agricultural land, but the fact that it has development rights enhances its value. So what is the starting value (X)? One method of establishing this is to look at what the cost (Y) will be to take that plot of land from its current condition to the more valuable developed state (Z).

As we know the value of the completed property from comparisons, and we can also establish the costs of the development from the estimates that will be necessary, we can work out the value left in the scheme (the residual) that can be applied to the land on its own.

Put simply, X (the value of the land) = Z (the value of the completed development) − Y (the costs of the development)

Example: value of completed development	= £500,000 (Z)
less development costs	= £320,000 (Y)
equals land value	= £180,000 (X)

X becomes the residual value. Hence the reason for the method being called a residual valuation. This example is in its very simplest form and understanding how the costs are broken down and the need for finance costs, developers' profits, professional fees and the application of taxation all add another dimension to the thought process. In summary, the land value is the completed value *less* what it takes to get it to that point.

This is the traditional way that developers establish the value of land, by starting with the sale price of the properties that they intend to build and working backwards to understand the value of the land to them. This is done by using the comparable valuation approach to value the completed development. This gives the gross value of the completed development. All the costs associated with the development process are then deducted from the end value, leaving the remainder, which is the price that the developer should be able to pay for the basic site. However, when looking at a large site, it becomes much more complex, as time has to be factored in, which includes the cost of finance over an expanded timescale together with phased sales and cash flow, which cause the actual costs to be multiplied significantly.

We will not go into this level of detail in this book, as the focus here is for a single home development. Principally, what should be included in the costs and how these can be verified?

Costs of development include the costs of getting planning and building regulation approval, which for a single development may be more expensive in proportion to the larger site, where there could be economies of scale. This principle applies to all parts of this process.

The full list of costs is likely to include:

- regulatory fees (planning and building control requirements);
- professional fees (such as site acquisition, provision of plans, support with planners, sales costs, a quantity surveyor to cost the scheme, project manager);

- ground and external works;
- build costs (materials, labour and supervision);
- contingencies;
- funding costs (interest) and tax, such as VAT;
- sale and purchase costs in addition to fees that may include Stamp Duty Land Tax;
- developer's profit (where this is relevant).

Each scheme is likely to be different; in some cases it may not be necessary to employ a project manager, whereas in others it could be essential. In a self-build scheme there may be no profit factored in, however, if the scheme is funded by a lender and that lender had to take over the scheme (in possession), then it would be necessary to factor in a developer's profit.

Taxation, such as VAT, can also be treated differently and tax rates vary over time, so we cannot be precise here. The HMRC website (www.gov.uk/government/organisations/hm-revenue-customs) is the place to find the current rates and exemptions. For example, there may be provision for a full refund of VAT on building materials, and work on property for someone such as a disabled person or with environmentally friendly materials may qualify for a reduced VAT rate. Detailed advice on VAT and property transactions can be found in the Inspectors' Manuals (www.gov.uk/hmrc-internal-manuals/vat-land-and-property).

10.2.1 *Key factors in undertaking a Residual valuation*

Scope   The importance in each case of looking at the individual circumstances and not being governed by the limitations of a normal mortgage valuation, other than in the extent of inspection, definition of value and key assumptions. Therefore, make sure that you are getting a fee that reflects the work necessary to complete the job correctly.

Experience   If you are going to undertake this type of work, then good experience of building practice and procedures with knowledge of costs is useful.

Research   Your research in these cases needs to be comprehensive; look at the market demand and any issues on the horizon. This is going to be a transaction that stretches over a longer period of time than normal. As a new plot, are the boundaries defined? Are there any rights of way? What is the situation with drainage and provision of other utilities? Is there a risk of flooding? And what other environmental issues might affect this new build?

Report format   Provide a report format that allows you to include all the information that could materially influence the value. Be transparent with any issues and special conditions that may need to be applied to ensure this development will complete in the way anticipated, in line with budget and on time.

Regulation   Check the planning and building consents, looking very carefully at any conditions, especially as to future use. It is quite common, for example, to find that rural plots were given planning consent only for the use of those currently or previously employed in agriculture or similar. Also be alert to any restrictive covenants which might impact development, and which take precedence over planning permission.

Costs   Make sure there is a good quotation for costs, supported by a qualified person. The reality is that it could be an estimate from a builder with no provision for contingencies and no end date. This is where your expertise has to come in, not necessarily in determining costs, although that would help, but in knowing who to consult to confirm the proposed development financials.

Valuation   Being able to visualise the finished development is also going to help, as in most cases money is tight, especially in the early to mid-stages of development, when the costs may exceed a current value and pressure may be applied to achieve a good final figure. Anticipating the specification with unknown builders is another challenge that needs to be reflected in the assumptions and made abundantly clear to all those involved.

Supervision   It is unlikely that the lender will want you to go back during the inspection, and supervision would be beyond the scope of your instruction, but it poses a significant area of uncertainty and therefore, architect supervision or some warranty scheme might help with this area.

More detailed examples of this approach are given in works such as *Property Valuation Techniques* by David Isaac and John O'Leary (2013).

10.3 The Investment approach to valuation

The income approach to valuation, or investment valuation, is the method by which a value can be placed on an income-producing asset or property. In this case, the value is derived from converting future cash flows, or income, into a single current value.

The principle of this approach is that the motivation for buy-to-let investors in residential property is different from that which drives a value in the owner-occupier market. In these cases, the property is being purchased solely as an income-producing investment, with the potential for capital growth over time. This does not mean that all buy-to-let purchases are valued in this way. A lot of it is down to the wishes of your client and also the availability of comparable evidence for properties in a similar use in the area. For standard single-family occupied rental properties, the comparable method is generally used, but as the occupation becomes more complex and we enter the HMO (Houses in Multiple Occupation) market, a valuer should consider if capitalising the income flow is a more appropriate approach.

An understanding of how relative investments perform is a useful starting point here and Savills' *UK Cross Sector Outlook* (Sparrow *et al.*, 2020) is a study that undertook a comparison for the start of the new decade on returns that may be expected for different types of property (see Figure 10.1). It is important to read the footnotes regarding how these figures were put together as, regardless of comments throughout this book about improvements in data, it was not easy for the precise returns to be found.

Figure 10.1 is demonstrating a five-year forecast based on past performance, in respect of rental (income) returns and the growth in capital value. This should not be used as an absolute indication of a current return, partly because it is a forecast, and, second, because it is, at the very least, regional, so regions will differ and locations within regions will differ. This is highlighted particularly by the very good annual return forecast for buy-to-let properties in the North West compared to the reasonable return nationally, but the lesser return in London.

Also note that forestry, nationally, is not anticipated to produce any annual returns but does give a good return for capital growth, so different investors will have different

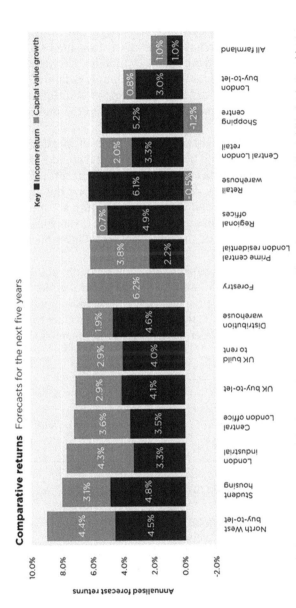

Figure 10.1 Comparative returns for differing types of property

aspirations on whether they need annual returns or capital growth or both. In addition, investment in certain sectors, such as forestry and agriculture, are significantly impacted by wider factors, such as taxation and subsidies (and these are not static).

What figures such as this provide is a benchmark at a point in time and, if it comes to the time when the valuer is doing a valuation and there is a variation from the benchmark, then a certain amount of investigation is warranted to establish what is making the difference. So, it could be that buy-to-let property in the North West is a good investment currently, as prices to purchase are relatively low, and yet the rental returns are relatively high. However, by applying simple economics, if investors increase their demand in the North West, then prices will rise, as supply is limited, and the annual returns are likely to fall because prices have gone up. In addition, there could then be less annual capital growth as demand decreases due to the high prices and lower returns.

The principle behind the investment approach is that an investor has a choice of where to place their capital. This could be in property, but it could equally be in shares, gilts or other traded assets. A core data comparison for investments is government bond yields. The returns on these will vary depending on how long the investor decides to invest so in January 2020 *The Financial Times* was reporting a 1-month rate for UK bonds of 0.73 per cent and a 30-year rate of 1.25 per cent. Investment in a UK government bond would be seen as one of the safest/least risky investments at the time of writing.

The decision on the ultimate investment vehicle is generally guided by comparing the perceived risks of each asset class, and the potential for a good return on money invested. This return is expressed as a percentage of the capital invested or the 'yield'. As a very simplified illustration, if £100,000 is invested and produces a return of £5,000 pa, this investment has a yield of 5 per cent, which is an expression of the risks of investing in that asset.

 Benchmark: Yields

Yield is the return on an investment. Generally, the higher the yield, the higher the risks associated with the investment.

If we are to show the yield as a simple mathematical calculation it would be:

$$\frac{\text{annual income}}{\text{present value}} \times 100 = \text{yield}$$

If this basic formula is applied to completed transactions in the market, a valuer can quickly establish the prevailing yield in the market and, having done this, apply it to the valuation of the property in question by converting the income flow (by means of rent) into a present value:

$$\frac{\text{annual income} \times 100}{\text{yield}} = \text{present value}$$

For example, if a property is achieving a rental income of £15,000 pa and we have established that the prevailing yield in the area is at 5 per cent, the present value is £300,000.

$$\frac{£15,000 \times 100}{5} = £300,000$$

In this context the 100/5 expressed in this calculation is also known as the 'years purchase' (YP), or 'present value of £1' (PV£). In simple terms, how many years it will take you to get your initial investment back. In the example above, the YP, or PV£ is 20 years. The higher the yield is, the shorter the time it will take for the investment to pay back, so the lower the YP.

The explanation and example given above are, as you would expect, grossly simplified, not least because there are costs in investing in property and a basic (gross) yield does not reflect these. If the assets being compared and analysed are of a similar nature, this should be sufficient. However, for more complex analysis of different investment types, a 'net yield', or 'return on investment', which takes into account costs, will need to be used.

In the property context, the costs of owning and managing an HMO property are significantly greater than a property for single-family occupation, so comparison of gross yields on these properties would not be appropriate. To come up with a net yield, the valuer needs to consider and deduct items such as insurance costs, management charges, maintenance, licensing, a sinking fund for repairs, the costs of borrowing, etc. As a simplification, in these circumstances, an 'all risks yield' may be used as a single reflection of the costs of ownership. In contrast, a government bond has very limited cost attached to it – possibly just some brokerage cost for the administration.

For taxation: *capital growth* from bonds or shares will be treated in a similar way for tax purposes as the gains from other capital assets, such as certain real property. The taxation of *income* from shares has, for some time, been preferential to that from other investments and from rental income (and income from employment or a business), in that it is taxed at a significantly lower rate.

A factor which differs between investment valuation and the valuation of owner-occupied property is how to deal with condition. Where a tenant is involved with a residential property, then the influence of the Housing Acts must be considered. If there are defects, then these could render a property unsuitable for letting and until such time as they are rectified, the property could be vacant with no income. This is clearly something an investor does not want, and that a lender will be concerned about as regards getting future repayments of the mortgage. The Housing Acts cover a multiplicity of items, in fact, 29 are considered under the Housing Health and Safety Rating System (HHSRS), which is a risk-based evaluation tool to help local authorities identify and protect against potential risks and hazards to health and safety from any deficiencies identified in dwellings. It was introduced under the Housing Act 2004 and applies to residential properties in England and Wales.

It essentially relates to deficiencies and these are defined as: 'A failure to meet the ideal … The failure could be inherent, such as a result of the original design, construction or manufacture, or it could be a result of deterioration, disrepair or a lack of repair or maintenance' (Housing Health and Safety Rating System, *Operating Guidance*, Office of the Deputy Prime Minister).

This has wide-ranging impacts. A steep staircase on a property for owner occupation takes on more significance if the property is to be let, as it may make the property totally unsuitable for an elderly tenant or a regular visitor of that tenant. The impact is to reduce the letability of the property and therefore the rent it could command, or possibly an increased level of voids in seeking a new suitable tenant.

Any complaints of damp may be subject to enforcement procedure by the local housing authority, if the landlord does not find a quick solution. This obviously makes the owning of a property as an investment more onerous than one for owner occupation and this needs to be reflected in the assessment of risk. Standard comparable methods do not achieve this, but an investment method of valuation does.

10.3.1 *Residential investment as buy-to-let categories*

The buy-to-let market in the UK has seen a huge increase in demand in the last 20 years, as private investors have increasingly looked to the property market as a way to secure an income stream and a higher return on their investment than can be found from other sources. This rise in demand has brought with it both opportunities and challenges. These have not escaped the notice of successive governments, who have increasingly looked to regulate and control this sector. Any surveyor who is instructed to value an investment property, be it for individual family occupation or as a house in multiple occupation (HMO), must be fully up to date on the regulations, requirements, obligations and legislation that will have an impact on their valuation and the advice that they give to their client. As a result, valuation in this sector of the market is now considered to be a specialist field that should not be attempted without a full understanding of all of the drivers of value.

Categories of buy-to-let properties

The RICS Red Book categorises residential investment properties into three groups:

- properties let to a single household on an assured shorthold tenancy (AST);
- properties let on a single AST, but to a maximum of four individuals who are sharing;
- houses in multiple occupation (HMO) and multiple units on the same title.

In addition to these categories, there are still some properties that would be subject to regulated rents. These are rents that date back to the years when rent control was applied and define certain types of tenancies. They will be relatively rare, but it is important that before undertaking such a valuation that the terms of the tenancy are established, and an income approach applied to the valuation methodology. However, some buyers will look at the anticipation of getting vacant possession, which may be the life expectancy of the tenant, and discount the capital value. These valuations should not be attempted without available evidence of similar market transactions.

10.3.2 Receipt of instruction

On receipt of an instruction to value a residential buy-to-let property, it is important to establish the following facts:

- the category of the property (see above);
- the basis on which it is to be rented;
- the valuation approach required by the client. This may be left to the discretion of the valuer, but many clients are specific on how the valuation should be completed.
- the location of the property. Is it in an area of high rental occupancy? Is selective licensing applicable? Is there an Article 4 Direction applying in the area?
- if the property is leasehold, the main terms of the lease.

10.3.3 Valuation of single-family occupancy properties

The valuation of a standard buy-to-let property that will be used for single-family occupation is very much the same as that for owner-occupied housing, with lenders and purchasers in this sector looking at a value based on comparable sales of similar properties (see Chapter 6). In choosing your comparables, it is important to understand the nature of the location for both the subject and comparable properties you have chosen. Factors such as demand for rental properties can be very localised and driven by the sources of tenants and the controls imposed on renting in any specific area. It is worth bearing in mind that some lender clients will not want to lend on a buy-to-let property in an area where there is no demand for owner-occupied properties; for this reason (as much as any other) it is always important to refer to your client's guidance or policies prior to valuing and recommending a property as suitable security.

Whilst the extent of regulation impacting single-family occupation properties is lower than for HMO, any valuer must be fully aware of the requirements of the Homes (Fitness for Habitation) Act 2018 and the statutory licensing requirements.

10.3.4 Valuation of Homes in Multiple Occupation (HMOs)

The valuation of HMO properties is significantly more complex than mainstream single-family buy-to-let properties, as this sector is much more heavily regulated, and the requirements and obligations change frequently. A full and up-to-date knowledge of the legislation and regulation that apply to this sector is a critical requirement before a valuation is even attempted.

Current regulation includes:

Housing Act 2004: This Act defines an HMO which, at a high level, is a property occupied by three or more tenants from two separate households, who share a kitchen, bathroom or toilet.

Management of Houses in Multiple Occupation (England) Regulations 2006: These regulations place obligations on landlords for the maintenance of the building and services. They set safety requirements, minimum lettable room sizes, obligations for the disposal of waste and the also for the provision of information to tenants.

Housing Health and Safety Rating System: This is the basis of assessment used by local authorities when considering rental properties (particularly HMOs). It uses a risk-based approach which considers different elements/hazards and their impact on the occupants, based on their vulnerability.

Fire safety legislation: There have been several pieces of legislation that have looked to address the increasing concern regarding fire safety in buildings. The most recent of these is the Fire Safety Act 2021.

Energy Act 2011: This introduced the minimum energy-efficiency standards (MEES), which progressively restricts the ability to rent out a property which does not meet the required energy-efficiency threshold (exceptions apply).

Licensing of HMOs: The Housing Act 2004 (amended 2018) laid down mandatory requirements for HMO properties to be licensed. This obligation currently applies to properties with five or more occupiers and will set down a limit on the number of occupiers allowed. Each licence is personal to the owner of the property and is not passed on with a sale.

Selective licensing schemes are implemented by individual local authority areas and have the effect of bringing additional properties under the licensing umbrella. Selective licensing is usually imposed in areas of social deprivation, or where there are a high number of rental properties. A valuer must always be familiar with areas in their locality which are impacted by these obligations, as they will affect the ability of a property owner to rent out their house or flat and so will need to be reflected in value and mortgageability.

Article 4 Direction: An Article 4 Direction is a planning control that was introduced under article 4 of the Town and Country Planning (General Permitted Development) Order 1995. The effect of this control is to remove all permitted development rights in a particular defined area. This means that planning consent is required for any change of use to a property. There is no automatic right to convert a single use residential unit (planning category C3) to a small HMO (planning category C4).

All of the above mean that a thorough knowledge of the location, the local rental and buy-to-let market, regulations (both local and national) and legislation is a prerequisite to undertaking the valuation of an HMO property. All of these will impact on the ability to rent out a property and thus its value and suitability for a buy-to-let mortgage.

When valuing HMO properties, the nature of the unit and your client's instructions will drive the approach to valuation. In some cases, such as larger HMOs and multi-unit properties, consideration should be given to the appropriateness of the investment approach to valuation. In this event, the market rent will be capitalised at an appropriate yield (see Chapter 9). The type (gross or net) and level of yield chosen will vary according to the nature of the unit being valued and it is important that, in using this method, the valuer takes care to fully justify their approach and yield choices. In addition, when using the investment approach, it is always wise to undertake an additional sense check of the investment calculation outcome, by comparing this against recent sales evidence of similar HMO rental properties in the area.

All buy-to-let instructions will require the provision of a rental valuation, which should be provided in line with the Red Book definition and defined assumptions. As with any valuation, it is wise to compare like with like and to factor in such considerations as running costs, repair works and management charges.

10.3.5 Alternative income approach valuation methods

As a final short reference in this section, it is worth mentioning that the income approach to valuation also covers a **discounted cash flow** approach, which discounts all future cash flows to a present-day value, and also the **term and reversion method**, which is used where the property income is known for a fixed period, but that it will then be reviewed at set times in the future. This income is capitalised at different yields and brought back to a present-day value. These approaches are dealt with in more detail in specialist commercial property valuation textbooks.

As a sense check for those residential valuers undertaking a simple investment approach valuation, the best approach is as follows:

1. Establish comparable sales of similar properties.
2. Find rental incomes for these properties (you may need to talk to local agents).
3. Use this information to establish the prevailing yield in this market. Adjust to reflect the subject property, if necessary.
4. Capitalise the rental income of the subject property, using the established yield.
5. Analyse in a matrix.
6. Document your analysis and thought processes.
7. Cross-check with a comparable analysis of value to see if the conclusions are supported by an appropriate rationale.

10.4 Leasehold properties

Leasehold properties hold their own particular place in the residential property market, and for many years (particularly outside London), little regard was had for the impact that this tenure can have on value.

In essence, when a person buys a leasehold property, they are paying for the right to occupy or 'enjoy' that property for a fixed period of years. At the end of this period the property will revert back to the freeholder. For this reason, an interest in a leasehold property is, by nature, a diminishing asset; it stands to reason that the right to occupy a property for 125 years will be worth more than the right to occupy it for only 25 years. It is against this backdrop that we need to understand the drivers behind values in this sector of the market.

To the average uninitiated purchaser a leasehold property with 65 years remaining on the lease is a perfectly good buy. After all, they probably only want to live there for a few years, so may see no reason to offer a lower figure for this interest. However, from a lender perspective, if the purchaser takes out a 25-year mortgage, at the end of the mortgage term the lease will only have 40 years to run, and any future lender will be reluctant to take this on as a risk.

To address these issues there has been legislation over the years, starting with the Leasehold Reform Act 1967, which gave occupiers of leasehold houses the right to extend their lease, or 'enfranchise' (i.e. collectively purchase the freehold). This was followed more recently by the Leasehold Reform, Housing and Urban Development Act 1993, which gave the rights of collective enfranchisement for blocks of flats, or the right to a new lease for flat leaseholders (90 years at a peppercorn ground rent). These Acts also laid down the process by which this would take place and the principles of valuation that should be followed in calculating the compensation to be paid to the freeholder in granting these rights.

Valuation for leasehold reform purposes is a separate discipline in its own right and is not covered in detail here; however, simply put, the principle behind the calculation is that the compensation paid to the freeholder is the aggregate of:

- the diminution in value of the freehold interest;
- 50 per cent of the 'marriage value'; and
- compensation for losses (costs, fees, etc.).

In order to exercise these rights, the leaseholder must hold a lease for an initial term of more than 21 years and must have been in occupation for more than 2 years (though once notice to exercise the right has been served, this can be assigned to any purchaser of an existing lease).

The term 'marriage value' may be new to students, but can be defined as the 'value released by the coalescence of the freehold and leasehold interests'. In order to calculate this, the leasehold reform valuer will work out the value of the aggregate of all the interests prior to the transaction and deduct these from the value of the aggregate of all the interests once the transaction has taken place. Thus 50 per cent of this value is payable to the freeholder, though it is also worth noting that, under the Act, this payment is deemed to be £0 where there are greater than 80 years remaining on the original lease term.

Benchmark: Marriage Value

The marriage value is the value released by the coalescence of the freehold and leasehold interests. This only becomes payable when there are 80 or fewer years remaining on the lease.

Based on the above, it stands to reason that the costs of applying for a lease extension will increase as the remaining term of the lease gets shorter and that these costs will increase significantly once the 80-year threshold is passed.

Without a full calculation of costs, it is hard to understand the full potential impact on the value of the property in the re-sale market. However, over the years a number of practitioners (and the RICS) have developed graphs that depict the relationship of the value of the leasehold interest against the value of the freehold interest at any given point in the lease, as well as the relative value of the property in relation to its unexpired term. These 'relativity graphs' can be a useful tool in assisting in the calculation of the value of a leasehold property.

Benchmark: Relativity

The relative value of a property that is held on an existing long lease, as compared to its value as a freehold.

In *Mundy v Trustees of the Sloane Stanley Estate (2018)* the Court of Appeal rejected a computer model used in valuing lease claims when it resulted in an impossible outcome, highlighting that graphs and models must be reviewed for specific applicability and, not least, common-sense.

Caution should, however, be employed in using these graphs, as they have their limitations. Whilst they all follow the same general path, they are very subjective, being based on tribunal decisions and market experience from a limited number of cases. They also employ different assumptions from each other (such as a 'no Act' world) and are very specific to geographic areas. The example in Figure 10.2 shows a representation of how the relativity would appear for just one supplier of the information. Quite often a number of curves can be shown on the same graph and this demonstrates that there are variances between the versions produced from different sources. These graphs should not be relied on as a valuation tool without an understanding of their limitations and without supporting evidence, or a sense check for your conclusion being provided by an additional other means of analysis. The graph shows that the longer the remainder of the lease, then the less impact on value, but with only 50 years remaining, there is a fall from approximately 80 per cent of full freehold value down to zero shortly before the lease expires, when the leaseholder would vacate the property.

So, at a high level, the remaining number of years on a lease is a major value driver in this market. This in itself presents challenges to the valuer, but the influencers of value in this sector are greater than this, as there is an array of other elements that need to be understood if you are to arrive at a reasoned conclusion of value. The following

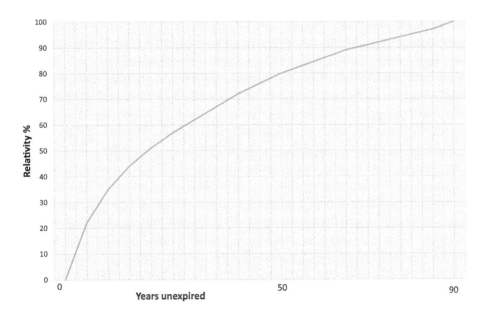

Figure 10.2 A representation of a relativity graph. There are many versions available which may be based on different geographic locations or different assumptions. Take care in using these, they are a supporting tool and not an answer in themselves.

list is an indication of the other elements that must be considered, but cannot be taken as exhaustive. There are an infinite number of lease models, all with different clauses, some of which may not be reasonable. In addition, continuing government interference in this sector may serve to distort values and further influence your decision process in arriving at a valuation figure.

At the very least, the following matters must be considered as potential value impactors:

1. **The terms of the ground rent**: Is the starting/existing level acceptable to mortgage lenders? What are the review periods? What is the escalation clause?
2. **Service charges**: Are they reasonable? Do they reflect the services provided? Importantly, what is the escalation mechanism?
3. **Rentcharges**: Are these reasonable? What is the escalation mechanism?
4. **Event fees**: Do these exist? Are they reasonable?
5. **Foreseeable costs:**. Are any large costs anticipated that may need to be apportioned amongst the leaseholders as a one-off charge? This is particularly pertinent on properties where there is no sinking fund mechanism in place to cover such costs.
6. **Maintenance**: Who is the freeholder? Are adequate maintenance arrangements in place? Is the building being maintained?

For many years, valuers in this market relied on standard assumptions that were laid down in the RICS Red Book. This is no longer considered acceptable and, at the very least, the valuer is expected to make reasonable verbal enquiries about the nature of the lease with the occupier or agent. However, with the wide availability of data, the establishment of these facts prior to a valuation will become a matter of standard due diligence and competence and must be taken seriously in the provision of a professional service and the protection of clients' interests.

One final short note that is worth considering concerns leasehold properties in the rental market. It is important that you know your local market in these areas, as a prevalence of buy-to-let purchasers in any market can distort leasehold values, particularly where there are relatively short lease terms remaining. In some cases these investment purchasers will be prepared to target short lease terms, and pay slightly more for them, in the pursuit of higher returns.

In conclusion, whilst the valuation of property for leasehold purposes is outside the remit of the standard residential valuer, it is important that practitioners have a high level of understanding of this and the other complex value drivers in this sector.

10.5 New build property valuation

Whilst the basic principles for the valuation of new build properties are the same as those for any other residential unit, this sector of the market does bring with it additional complexities that need to be considered.

For the purposes of valuation the definition of new build varies slightly, but it is generally agreed that it covers properties that are newly constructed and that are to be occupied for the first time. For many lenders this also includes properties up to two years old and conversion projects where the property is to be occupied for the first time in its new form.

Personal preferences of buyers play a big part in the pricing and marketing of new properties. In the same way that anything new usually commands a higher value than something second-hand, property is no exception. Possibly the best parallel is a new car. When you drive it off the forecourt, then it usually drops in value because it is no longer new. However, unlike a new car, which usually deteriorates in value with age (unless it is a classic), a property may actually increase in value due to inflation and possibly presence in the marketplace.

In approaching such a valuation, the standard processes outlined above need to be adopted, but the following elements also need to be considered, and it is these that differentiate the new build from older properties and need to be recognised over and above the first occupation benefit:

- **Warranties**: New build properties generally come with a warranty. These usually cover minor defects for the first two years and serious defects for up to 10 years. The existence of this warranty is almost always a requirement for lender clients and, as such, the absence of a warranty will impact on the saleability, and thus value, of the property. For individual or self-build projects and small-scale developments, an alternative in the form of a professional consultant's certificate (PCC) is generally considered acceptable. Note that where warranties are contractual, they will only bind the parties and future purchasers will need to ensure the presence of wider guarantees, assignment of warranties or a PCC.
- **Sales approach**: The way new properties are sold is very different from the second-hand market. Rather than an individual sale through an estate agent, these properties are often widely exposed to the market through high profile marketing campaigns covering the whole development and a range of properties available, instigated by the developer.
- **Incentives**: Linked to the marketing approach of new build units, and the fact that a number of similar units are made available at the same time, the use of incentives to purchase is a well-established practice in this sector. The form that these incentives take is many and varied and it is the job of the valuer to see through these in order to understand the underlying base value of the property. In the past, the existence of incentives was often not disclosed to the valuer and this lack of transparency created challenges in the market. To address this, the Disclosure of Incentives Form (DIF) was released by CML (now UK Finance) in 2008. This form should be made available for all new build properties and on all large build schemes to identify key features of the development.
- **Government initiatives**: Over the years, successive governments have introduced schemes to encourage housebuilders and to make home ownership more affordable. Schemes such as shared ownership, shared equity, first homes, and Help to-Buy have made housing more affordable to many, but have introduced potential distortions in the market which the valuer must understand and interpret.

10.5.1 Drivers of value in new build houses

All new build properties, like houses in the second-hand market, are valued to market value, as defined by the Red Book, and as described for valuations throughout this chapter. As with second-hand properties, the drivers of value in the new build sector are determined by market forces. However, there is no doubt that new build properties bring with them additional attractions. Not least of these is the fact that no one has lived in the house before, paired with the availability of a customised specification and certainty of completion date. As with any other product, any value attributed to these elements is lost once the building is occupied and enters the second-hand market. This attraction creates what you will hear referred to as the 'newness premium', 'first occupier benefit' or 'new build premium'. It is in the challenge of establishing the extent to which this premium has added to the sale price that much of the complexity of new build property valuation lies.

In establishing the first occupier benefit, the valuer has to consider how much a purchaser is prepared to pay, above the price of a second-hand property, to get the benefit of being the first occupier of the house.

! Benchmark: First occupier benefit or new build premium

The uplift in value attributable to the 'newness' of a property. This uplift is lost once the building has been occupied.

A distinction needs to be made at this point between the newness uplift in value, that will disappear on re-sale, and the other attractions that are offered by a new property that will still hold value when the property is offered on the second-hand market. These will be elements such as innovative forms of construction, better use of space, energy efficiency, and adaptation for modern technology plus, in the early years, there is usually less maintenance required.

Finally, we have to consider any schemes through which the property is being purchased. Where a government or public sector body is offering assistance with purchase through loans, discounts or shared ownership, the impact of these schemes on the value, created by the increased levels of affordability, will need to be factored in. At worst, this effect will inflate the initial purchase price and then disappear on re-sale in the second-hand market. For some schemes, such as shared ownership or discounts with public sector-imposed occupancy restrictions, the affordability factor will remain on re-sale, but may be removed if the property is taken into possession by a lender during the term of the mortgage. It is for this reason that lenders ask for shared ownership properties to be valued as if they are offered as 100 per cent ownership on the open market, and properties offered under a one-off discount scheme may face a revised valuation as their sale price does not always stand scrutiny against comparable transactions in the second-hand market. More consideration of this is given in the comparable valuation section.

10.5.2 Inspection of new build properties

New build properties present inspection challenges all of their own, principally because, for health and safety reasons, an on-site inspection is generally not permitted. In this event, the valuer has to rely on site plans (construction site plans should be used, not marketing documents) combined with an inspection of a similar show home to get the feel of the finish and layout. Where this is the case, it should be fully documented in the site notes and care should be taken to ensure that the correct property is identified.

Platform Funding v Bank of Scotland (2012). In this case, liability was found when a surveyor valued the wrong property. He was fraudulently directed to the incorrect property, but the finding was not in negligence (where there could be a defence of 'reasonable' care having been taken), but in contract, due to an express term vouching that the property offered for security was the one inspected. Note then, that not only reasonable care and skill in ascertaining the correct property are required, but appropriate contract terms which do not unreasonably bind in the face of third part deception.

Where inspection is not possible, the Red Book allows a number of assumptions to be made, principally that no hazardous materials have been used in the construction. An additional special assumption routinely used with new build properties is that the construction is complete. This allows you to give a completed valuation on which the lender can base their lending.

Whilst on site, the valuer should endeavour to obtain a copy of the UKF DIF form for the subject property, as this will be a requirement prior to lending and will detail all the appropriate elements of the transaction which could influence the final valuation conclusion.

10.5.3 Comparable evidence

As with all valuations, comparable evidence on which to base a valuation decision will need to be sourced. In the new build sector it is important that the evidence used is not solely restricted to new build properties and that sales evidence from properties in the local second-hand market is also used.

Best practice suggests that the use of five comparable properties should cover all of the requirements. These should be taken from a range of similar new build properties on the same, or nearby, sites and from the re-sale market for properties of a similar size and, ideally, age. In analysing the evidence, the valuer needs to understand as much detail about the transactions (such as incentives) as is possible and to analyse based on the published guidance from RICS.

10.5.4 Assessing value

In their guidance on the valuation of new build properties RICS lay down a particular five-point approach to analysing comparable evidence to arrive at a valuation of a new property. This starts with an underlying base value and works upwards:

1. Identify the underlying value of the property in the given market. This will be indicated by sales of comparable second-hand units.
2. Adjust to reflect the modern design and specification. The level of adjustment here, if any, will depend on the nature of the comparable evidence used.

3. Factor in any incentives. How have these influenced the final sale figure? This is a value judgement and is not just an exercise in the deduction of the value of the incentives from the sale price – the two are not necessarily related.
4. Reflect the first occupier benefit. This will vary from location to location and with different property types and markets. The key here is local knowledge. Indeed, in some markets there may be no uplift in value for new properties at all. In adjusting, you will need to document your rationale for the chosen adjustment, based on your local knowledge.
5. Reflect the affordability premium. This is again down to judgement and will need a rationale to support it.

In this section it is also worth noting some of the pitfalls that may be present in analysing any one transaction. These are many, but it is worth asking yourself questions such as:

- Is it the end of a quarter or the financial year? Builders often discount sale prices to achieve targets at these points of the year.
- Are there any undisclosed cash back deals?
- Has a part-exchange deal been agreed as a part of the transaction?
- If it is a buy-to-let purchase? Is there a guaranteed rent on offer?
- If the block is being marketed to the investor market, are the large lenders accepting it and is their exposure being limited? This can affect marketability and thus value.
- How much other development is ongoing in the area? Is there an oversupply?
- What is the mix of affordable housing and private sector housing? Are the two separated or 'pepper-potted'?
- Is the development attracting foreign investors, which can inflate values artificially?
- Is there anything about the transaction to suggest that there is an element of fraud or money laundering at play?
- Are there any estate charges associated with the development? If so, are the terms of this reasonable?
- Is the property freehold or leasehold? If it is leasehold, are there onerous terms which could impact on value and mortgageability?
- Are there any Planning Conditions, such as a section 106 notice that place limitations on sales or add onerous provisions to the site, as the developer may try and build additional costs into the prices?

Section 106 Agreements

This is a legal agreement, under s106 of the Town and Country Planning Act 1990, between local authority and developer which is attached to certain planning permissions to ensure the developer includes works of benefit to the local community and infrastructure. This might take the form of building works, such as gardens or roads, or might be a contribution to local services, such as schools.

- Are there any ground conditions that will add to the build cost, such as piling the foundations, or making provision for flood defences that the developer may seek to build into the pricing structure?

- Are the comparables you use from open market sales? Some developers will sell a property to their own company in the early stages of a development, to help to set a price benchmark.
- Is the property a part of a large regeneration scheme, with the sale price reflecting the anticipated uplift in values in the area once regeneration is complete? If so, how are you going to approach the early sales which will transact at prices that the current second-hand market does not support?

In assessing all of the above you will need to take a balanced approach. Avoid looking to just support the advised purchase price and build your valuation conclusion up, supporting it with a thoroughly documented rationale.

10.5.5 Valuation considerations

Table 10.1 summarises the considerations necessary to approach the valuation of a new build property.

To summarise, in algebraic terms, the valuation of a new build property:

An existing second-hand property of a certain type, style and size would be compared to another one of similar size, style and type in a similar neighbourhood/location and timescale: $X = X$

For a new property (Z):

Y = a premium add-on that expires on occupation
W = a value for newness that is retained in the core value
X = the value of a second-hand similar type, style and size in a similar location and a recent sale
T = the incentive (that is not repeatable on re-sale) provided by the developer to achieve a sale

Therefore, for the new property in its simplest form:

$$Z = X + Y + W - T$$

10.5.6 Permitted development rights

An increasing number of properties are entering the residential market after being converted from commercial units under permitted development rights. These buildings were not originally constructed for residential occupation and the quality of conversions varies significantly. Each needs to be considered on its own merits and many will present a number of additional considerations which will influence your final valuation conclusion.

Table 10.1 Valuation considerations

Type of initiative	Approach to value	Comment
First-time occupier benefit	Consider sales of similar size and type of modern property nearby and compare the new sale price of the subject property with the re-sale price of the comparable. Provided there are limited variations, then the difference could well be the premium value for new. In a market where demand is low and there is a lot of supply, there may be no scope for the premium	If a premium is recognised, then most valuers would think a lender would not wish to include it, because as soon as the mortgage is completed, and the property occupied, then it will drop in value. However, this is not the case. Most lenders expect the first-time buyer benefit to be included within the valuation, but the valuer should always identify the situation and report on it.
Developer incentive, e.g. payment of fees, part-exchange of the customer's old property, cash-back, etc.	The sort of incentive here should be included in the UK Finance DIF. The common factor is that none of these incentives are available on the re-sale of that property, so the value attributed to them should be deducted from the purchase price and the property valued relative to other comparable sales.	This could be a straight deduction of the value of the incentive with information provided in the report for the rationale. However, often this is not the case. For a part-exchange, this is not so easy, as how much value do you place on this sort of transaction when it is unlikely you will know the figures nor the benefit to the parties? In such cases it will be similar to the premium. Make comparisons of cases where no part-exchange took place and see what the outcome is.
Developer incentive, e.g. choice of kitchen units, turfing of lawn, garage, etc.	These incentives are physical in nature and will remain with the property and therefore should be reflected in the value	Normal new-build valuations reflect newness as a genuine differentiator
Government schemes, e.g. Help-to-Buy	Over the years governments have introduced various initiatives to help first-time-buyers get on the housing purchase ladder. The effect of these schemes is usually to reduce the money required to deposit for the purchase of a home or to reduce the repayments on a mortgage over the first few years. These mainly apply to new build properties, although various schemes exist.	If the schemes are such that all properties in the area have the benefit, then it is possible that no action is required other than to advise a lender and the buyer of the situation. If the scheme is acting as an incentive, then it needs to be handled as for incentives above.

Table 10.1 Cont.

Type of initiative	Approach to value	Comment
	Care needs to be taken to compare sites where such schemes are not being used or, even within sites some properties may not be eligible, so it is important to see if the scheme is acting as an incentive, which will not be available on re-sale	
Lender incentives	Banks have operated similar schemes as those of the developers in offering mortgages where the borrower only needs to find a 5 per cent deposit, or the repayments are discounted for a period of years, before reverting to the full rate. Banks do not differentiate between new and old properties and many valuers will not be aware of the incentives being offered. The reality is that these are so common that the whole market can be based on some form of incentive and therefore no allowance is made by the valuer.	This approach caused significant problems in the 2007/8 financial crisis, because the incentives were removed by the lenders and the market slumped during 2009. This was particularly in respect of first-time buyers, where mortgages requiring less than a 20 per cent deposit became extremely rare. As a result, first-time buyers previously requiring £5,000 deposit for a £100,000 house now had to find £20,000 and that could not happen overnight.
Shared ownership where the buyer acquires a percentage of the property, say, 25 per cent, and the developer or housing association owns the remaining 75 per cent, for which the buyer pays an equivalent rent	In these cases there is an incentive, because the property is more affordable. Instead of finding a deposit for 100 per cent, the buyer only needs to find a deposit for 25 per cent, so in the £100,000 example it is now 20 per cent of £25,000 which amounts to £5,000 as opposed to £20,000. The buyer still has a combination of mortgage and rental payments, but these are usually comparable to monthly repayments for the mortgage. To value such cases, then, compare to similar sized and types of property that have sold for 100 per cent of the market value. A lender will then only lend on 25 per cent of that value.	Given that these properties are incentivised, it is not uncommon to find that the purchase price is actually more than comparable sales of property without the shared ownership option. This has to be considered as a premium as it will not be repeated if the property was sold with 100 per cent occupation.

(continued)

Table 10.1 Cont.

Type of initiative	Approach to value	Comment
Affordable Homes, such as homes sold at a discount to first-time buyers or those in special employment, such as the NHS	Affordable Homes can either come with a condition that the limitation applied on the first purchase continues, or it may be removed. In the latter case the property no longer remains an Affordable Home If the condition applies, then it may also come with a valuation condition in that the homes have to be valued at a percentage of market value relative to their original discount	If a valuer is asked to provide a valuation on an Affordable Home, then it is critical to find what conditions apply and not just assume that the property can sell at 100 per cent market value, as this could be misleading

! Benchmark: Permitted Development Rights

The automatic grant of planning permission allowing building works or a change of use to take place without the formal approval process.

Principally, these units may have been constructed in a different way from more standard housing. The materials used may require higher levels of maintenance, there may be restrictions on the warranty cover and the adaptation of an existing layout may mean that some units suffer from a lack of natural light. More significant than all of these, however, may be the location of the conversion. Units on industrial estates and in commercial areas tend to be unattractive to the private purchaser and will inevitably almost exclusively attract investor demand. Where this is the case, many lenders will not consider such properties, or will limit their exposure. As marketability is a huge driver in current and future demand, these valuations should be approached with caution.

If in doubt, in all cases, refer to your client or to lender policy in order to help drive your conclusions.

10.5.7 Development potential

Although not necessarily new build, this is an opportune time to include reference to property that has the potential for development, either through total demolition and re-building, or for the provision of the development of part of the site.

There are numerous situations where there is a large property sitting on a good-sized plot that has been demolished and replaced with a block of flats or a number of houses. Alternatively a property has been marketed with the potential for a building plot within the garden. In situations where the valuer sees such potential then, unless the property has actual planning consent for such a development, the uplift in value this represents should be ignored for the purposes of a mortgage lending valuation, with an appropriate comment in the report detailing your approach.

If the valuation is for some form of tax calculation such as Inheritance Tax, for the Charities Act, or for release of equity to an elderly person, then the normal requirements of market valuation will be applicable, and the potential value uplift should be reflected.

10.5.8 *The influence of buy-to-let on new build values*

At the peak of the buy-to-let era it was not uncommon for developments, quite often flats, to be sold in blocks to landlords looking to let the properties. This could cause some concern to the current owners who had purchased for owner occupation and found that the whole nature of the development had changed. Where it did, then quite often the re-sale prices were significantly reduced. This sort of information now has to be released through the DIF and a valuer has to consider what impact this might have on the value of individual units.

10.6 Specialist valuations

In addition to valuations undertaken for mortgage purposes, there are quite a few valuations undertaken that fall under the same sort of definition, i.e. a valuation that reflects factors that materially affect value, which does not require a full survey at the outset but an inspection similar to a mortgage valuation is undertaken. The purposes of the valuations vary but may be subject to specific definitions, such as a valuation for Inheritance Tax. Therefore, we will briefly cover these to say that the inspection level will be similar to the mortgage valuation, but that has to be agreed with the client and the valuer must determine whether the purpose of the valuation can be achieved by that level of inspection.

There is no standard format set down for such reports, but they will need to comply with the global requirements of the Red Book, as amended by the UK Supplement, but again this has to be agreed in the terms of engagement. Many will be commissioned by a solicitor, who may already have standard terms.

Quite a few valuations fall under the remit of that imposed by an Expert Witness, such as a valuation for matrimonial dispute, and therefore effectively the client is the Court. The format of such reports is governed by RICS guidance relating to 'Surveyors acting as an Expert Witness' (RICS, 2014c) and at the time of going to print, the 4th edition was the latest version. We do not intend to cover that here.

What is important is knowing for such types of valuation:

- the purpose of the valuation;
- any refinements on the definitions of value;
- the scope of the inspection and reporting required to meet the purpose.

So, for example, in undertaking a valuation for a charity, then the Charities Act lays down certain requirements depending on whether you are required to act in the acquisition or the disposal of an asset. In this respect the Red Book UK national Supplement, issued in 2017, UK VPGA 8 is your core reference.

So, as an example, the reporting required amongst other things to advise on disposal should include measurements, lease details, whether it would be in the best interests of the charity to undertake repairs or whether there is any development potential. This is all to ensure that the best price is achieved for the charity. Therefore, the remit goes much further than a typical mortgage valuation.

Valuations for tax purposes also have differing criteria and again Information can be found in the Red Book UK Supplement for 2017 under UK VPGA 15, which covers Valuations for Capital Gains Tax, Inheritance Tax, Stamp Duty Land Tax and the Annual Tax on Enveloped Dwellings.

In such cases, there are numerous different requirements, and one is that the definition of Market Value is amended and includes the wording:

> the price which the property might reasonably be expected to fetch if sold in the open market at that time, but that price must not be assumed to be reduced on the grounds that the whole property is to be placed on the market at one and the same time.

The inference here is that larger estates where the option to subdivide the property may produce a higher value should be accounted for in taxation. For the typical mid-terraced house this is unlikely to be an option. However, knowing that valuation requirements differ is the starting point.

10.7 Valuing in periods of uncertainty

All valuations require an element of analysis and an understanding of matters relating to routine market risk. However, from time to time, major and unforeseen events will hit, and the impact of these on the property market is entirely unknown for a period of time.

RICS address the problems of valuing during times of market uncertainty in the Red Book. They take care to stress the difference between standard market risk and those times of uncertainty that make a valuer's task extremely challenging in interpreting the market to come up with a valuation conclusion. Examples of such events are the fall of Barings Bank in 2007, the surprise Brexit vote in 2016, or the outbreak of a global pandemic in 2020.

At such times a valuation is still required, but the amount of information available to the valuer is limited and the application of assumptions may be inappropriate or required to such an extent as to make the valuation worthless. The Red Book makes clear that, in these circumstances, the valuer has a duty to signpost the risk associated with their valuation conclusions in their report, with appropriate commentary. In this way the client is fully appraised of the risks associated with relying on your advice, but which lie outside your control.

A valuer will, however, still be expected to formulate some sort of valuation regardless of the situation, especially if required in order for a lender to release funds. In this event the approach to reaching a conclusion must be considered carefully.

Assuming there is little or no current market activity, then the benchmark must be the market and the comparables that were available before the situation arose. Clearly these do not now reflect the current situation, especially if it was unforeseen, but they are the starting point that can be supported by a rationale and a statement of material uncertainty that assumes the values will return to pre-event levels in the relatively short term.

However, this will very much depend on the scenario at the time. If the comparable evidence available at the time of the valuation indicates a longer-term decline in market values, then of course that should be factored in, unless there are special circumstances that have distorted their marketability.

It is not for the valuer to predict what may happen but look at what has happened and then qualify their current conclusion with a rationale of what is happening at the time of the valuation. The client should then have sufficient information to assess the risks associated with such a qualified valuation and can decide what mitigating action to take.

10.8 Technology and valuation

10.8.1 Overview

For many years, the valuation industry and its processes remained unchanged. However, in recent years, with advances in technology and the increasing availability of data, this is no longer the case. These advances not only relate to data, but to other elements of the job, such as advanced camera technology, allowing infra-red photographs to be taken to demonstrate issues with insulation or dampness. In addition, cameras on poles or on drones allow previously inaccessible areas to be inspected. As a result of this, it is no longer feasible to exclude such areas from even simple surveys, although some degree of specialism may be required to interpret the results. Such developments will mean that those areas of previous uncertainty now become much more meaningful.

In this section we are going to focus on the improvements in the use of data and how this could assist the valuer in producing a more reliable valuation.

A significant step forward in the accessibility of information came as a result of the computerisation of data, and the introduction of digital records for the Land Registry (HMLR) meant that most residential transactions became available for all valuers to use.

If you wish to understand more about what is not included in HMLR records and the processes behind the figures, then consult the website (www.gov.uk/government/organisations/land-registry). There are also many other sites that offer alternative information, but beware, as they are not the official government site and additional charges may apply.

Various companies have developed software to supply organisations, with variations on the data suited to their business needs. For valuers, the core software tends to supply a databank of property, searchable under various attributes, such as location, size and type. However, the data from the Land Registry is limited and does not, for example, record the difference between a house and a bungalow. To address this, the software suppliers have added supplemental information, such as from estate agencies, to fill in the gaps. This data is generally very good, but may not be 100 per cent accurate, so users need to be aware of its limitations, e.g. some agents may describe a property as a bungalow even though it has a room conversion in the roof and is thus not a true bungalow. The sophistication of these systems is such that the agent details will normally be attached together with numerous photographs, so the valuer can refine the results.

All of this relies on valuers working together to share their data in a co-operative way but, provided that they continue to do this, then the availability of this data now means that all valuers should have access to it and there is scope for a more consistent approach to reaching valuation conclusions.

Some organisations have gone further than this with their data, by seeking to emulate what the valuer does and producing an automated valuation model (AVM). Others, mainly statisticians, have developed mapping where value trends and buying preferences can be identified.

10.8.2 Recording of information

The collection of data is helped as more and more surveyors record their site notes electronically, using tablets or similar. The recording of information digitally still has to comply with the current standards for site notes, so it is only the methodology that changes. The significance here is that instead of the data being on a piece of paper, which was difficult to then record electronically, all the attributes and their nuances are now available to use any time and anywhere. Where this will lead is all speculation for the future. However, past research into such property indexation systems, like the Halifax house price index, did show that it was the core attributes, such as number of bedrooms, type of property and size, that had the greatest impact and many of the lesser attributes, such as a garage, had a marginal impact upon the core value.

10.8.3 Desk-top reviews and research

The precise use of this information was discussed in Chapter 8, but it is worthy of a mention here because it is part of the data collection that has revolutionised the information that both makes up the valuation and the customer feedback. We now have access to mining information, the geology (so we know what rock and soil formations are beneath the property), also whether an area is affected by radon gas or is likely to be flooded. The list is endless and therefore, rather than getting a surprise because a property is flooded, the buyer is better placed to make an informed decision and the valuer can therefore reflect that in the valuation. As technology has been employed, we have seen working practices alter and client requirements shift, such that the surveyor who does not keep fully abreast of the changes will find themselves at a competitive disadvantage or even on the wrong end of a negligence claim.

10.8.4 Desk-top valuations

As more and more information becomes available through the internet, the lenders have started to look to move lower-risk lending cases to a hybrid desk-top valuation approach. Such an approach has the benefits of the human skills of valuation, without a physical inspection. This brings with it the risks associated with valuing 'partially sighted', which include an inability to fully assess condition and the subtler nuances of location, but the benefits of reduced cost and a faster turnaround.

Any valuer who completes a desk-top valuation must weigh up the risks of doing so as a key element of accepting the instruction.

For low loan to value, lower-risk lending, there is no reason why a reasonably accurate value assessment cannot be made in this way, as long as the valuer has access to the following resources on the internet:

- The property is visible on street view, with a recent image available (within the last few years).
- Details of the property are available on comparable web sites, showing internal condition and layout, in the last few years. If the property is a standard unit with no particulars available, a valuation based on an agreed assumption of average condition presentation may be possible.
- The boundaries and extent of the property can be defined online.
- The property is not historic or an exceptional 'premium' house.

Where there are doubts about any element of the property, or your research suggests that the property may now be materially different from the house that is visible online (e.g. evidence suggests a recent extension, refurbishment or major alteration), the instruction to complete a desk-top valuation should be declined and your client advised that a full inspection is appropriate.

In agreeing to undertake desk-top valuations, both parties must be fully aware of the limitations of the advice given and in agreement on the acceptable parameters of this approach. This should be documented contractually.

10.9 What is an AVM?

Work practices aside, we have seen technological advancement being used by lenders, in particular, through automated valuation models (AVMs), which speed up the valuation process for lower risk properties. At the same time as bringing new opportunities, the advancement of technology has also put more focus on accessibility of the valuer's work leading to greater levels of audit scrutiny.

 Benchmark: Automated Valuation Models (AVMs)

The provision of automated valuations of property using mathematical modelling combined with a database. AVMs calculate a property's value at a specific point in time by analysing values of comparable properties on their database.

Since the early 2000s, AVMs have been undergoing development and have rapidly increased in accuracy, so much so that, for standard properties, they can now stand scrutiny alongside a human valuation. The outcome of this is that our lender clients are increasingly turning to this as a faster and cheaper route to valuation, particularly for their lower risk and less complex lending work. This is done through a risk-based approach, by setting rules within the requirements laid down by the regulators and with which the associated risks with an automated valuation are accepted.

So what is an AVM, what are the benefits and what are the weaknesses?

An AVM was defined in 2008 by the RICS AVM Standards Working Group:

Automated Valuation Models use one or more mathematical techniques to provide an estimate of value of a specified property at a specified date, accompanied by a measure of confidence in the accuracy of the result, without human intervention post-initiation.

AVMs can operate independently of any human intervention. They can also, however, be used to assist a qualified valuer in producing an estimate of value. It should be noted that, if used as such, the valuation will be subject to review and any necessary refinement in order to ensure that it complies with the standards in the Red Book, as this will no longer be solely the output of the AVM.

What the valuer needs to be aware of is that the modelling techniques can vary. As it is highly unlikely the developer will share their particular rules, it is only by trialling the performance of one system against another and by reference to human valuations that the reliability of the model can be checked.

More information on the types of models can be found in the RICS Information Paper on AVMs: www.isurv.com/.../automated_valuation_models_avms

10.9.1 Use of data

Whilst all AVM models will need to use the same attribute data as any valuer, it is important this data is accurate, as the impact of missing data on any one particular valuation can be significant. This risk is offset in large-scale lender valuations, because their model will use thousands of properties where a few errors will be balanced out or become irrelevant.

Locational criteria are also important and the base for this will usually be the post-code, along with socio-economic data based on the census data collected periodically. However, for the same reason as for de-sensitising errors, then there needs to be sufficient volume of sales within a postcode for the AVM to work effectively, so in rural areas where there is a diverse mix of property and a greater spread of attributes, then the AVM may have a lower confidence score. This is, in many ways, similar for a valuer who, with limited evidence to work with, will have to make more adjustments to achieve a valuation outcome. The difference here is that the AVM cannot make value adjustments in the same way, as it does not have the information and cannot make a subjective judgement.

Other areas for concern are where a property has changed since the data was collected and include:

- **Property alterations**: An example would be a large extension put onto a house recently, which may not be recorded within the data. As a result the AVM will work out a valuation based on the previous size.
- **Condition factors**: Any inclusion of condition data will, again, be at a point in time and, although age-related deterioration may well be reflected in the sale prices of similar styled and aged properties, more specific issues such as an outbreak of dry rot will not be reflected in the AVM.
- **Market factors**: A volatile market may also impact upon the ability of the AVM to perform, this could range from a shortage of sales, as happened in 2009/10, to a quickly rising market, where the collection of sold house price data is delayed by a few months and cannot be included in the modelling. In most cases, asking prices will also be included in the modelling, but they are susceptible to variation on completion, so this could build in some inaccuracy.

If faced with an automated valuation, consider if any of the above issues could apply to frustrate the outcome.

10.9.2 Mass use of data

Conversely, there are areas where an AVM can work very well, such as:

- **Urban areas** with a high volume of transactions of similar property (in the same way that a valuer would find it easier in such circumstances and be able to produce high confidence results).
- **Mass appraisal**, such as determining the value of a lender's mortgage portfolio.
- **Major urban redevelopment schemes** in assessing potential property acquisition costs assessment, e.g. HS1 and 2.
- **Taxation assessment** on property.

What started as a small-scale process ramped up in the 2010s onwards, to the extent that close to 50 per cent of valuations were being processed through AVMs as 2020 approached. As a result of this, valuers must seek to capitalise on the opportunities this provides to address complex valuations and to upskill and improve the status of the profession.

Those surveyors who have spent much of their career looking at standard estate houses, with no opportunity for progression, can look to the specialist valuation niches to provide opportunities to develop.

By specialist sectors, we can look to those parts of the residential market where government intervention has added complexity, in particular, the buy-to-let market, new build and leasehold properties. In these markets, increasing legislation and regulation have distorted values and, with the access to information that is now available to the valuer, interpretation of data to arrive at a value decision is an added value that the surveyor can give to lenders in helping them to understand and protect their investments.

In conclusion, the increased availability of information about individual properties may result in the need for fewer valuers, however, it also provides tremendous opportunities for those who can embrace the changes and see them for the opportunities that they provide. Whilst the fundamental principles behind valuing properties will not change, the skills required of the valuer in interpreting information *are* changing and there will always be a role for those who have a deep and thorough understanding of complex markets.

11 Additional legal matters

11.1 Overview

In Chapter 5, we encountered negligence, which is one of the major areas of tort law, i.e. a class of civil wrong. This chapter provides an overview of other areas of tort of which valuers should be aware: nuisance and trespass. This chapter also discusses other legal provisions of relevance to a valuer– health and safety law, bribery, money laundering, the law surrounding data protection, and also the framework of the legal structures of a business, whether a limited company, partnership or sole trader.

11.1.1 Nuisance

Private nuisance is an idea which will be familiar to readers, but the details of the law may be surprising. This area of law may be defined as an unreasonable interference with the *claimant's use or enjoyment of their land* as evidenced by either an encroachment (such as tree roots coming through from neighbouring property), direct physical damage (the tree roots causing building disturbance), or interference with quiet enjoyment (such as noise, smells, smoke, etc.).

Being listed under the heading of tort, it is clear that this is a civil wrong. Nuisance can, incidentally, also be a crime under the common law (per *Attorney General v P YA Quarries Ltd* (1957)) or, for example, under the Environmental Protection Act 1990, but that is beyond the scope of this book. We are looking at private nuisance as it affects residential property.

The indication of use or enjoyment of property shows that in order to bring an action in nuisance, the claimant must have a proprietary interest, i.e. be the freeholder or leaseholder of property. Although we may think of it as a personal problem (when noise or fumes are causing annoyance or even damage to health), it is only actionable in the context of unreasonable interference with the use of property as established in *Malone v Laskey* (1907) and reinforced by *Hunter v Canary Wharf* (1997).

Note the requirement that interference is *unreasonable*. This will vary depending on the nature of the area (per *Sturges v Bridgman* (1879)) and reasonable expectations in that context. It is unlikely, although not impossible, that a one-off event be considered a nuisance. But the widely held view that if one comes to a nuisance, one cannot sue is incorrect, as underlined in *Coventry v Lawrence* (2015). If a nuisance has gone on for a long period and the claimant is new to the area, that might be part of the equation as to deciding reasonableness, but it is not definitive in permitting something to continue which is, objectively, a nuisance.

DOI: 10.1201/9780367816988-11

A final point to note on private nuisance is that there is no need to establish a lack of care. There will often, in fact, be a lack of care and the scenario may well be negligence as well as a nuisance. But lack of care, whether subjective or objective, is not a requirement.

11.1.2 Trespass

As property professionals, there may be a tendency to suppose that trespass is purely a land-based wrong. It is land-based but, beyond the scope of this book, there are the related wrongs of trespass to person and trespass to personal property. In short, trespass is doing something without right and, in our context, going onto land, or causing something else (a tree branch, rubbish, goods of any sort including livestock) to go onto another's land. To be liable you need to know it is another's land where you have no right to be, be careless (negligent) of the fact or remain there after you find out.

Trespass is, for the most part, a civil wrong which means that the remedy, where personal request does not work, is through the County Court to obtain a possession order if people or their possessions are still on your land, or to obtain an injunction to stop them coming onto your land. This explains why signs on buildings, fences and farm gates indicating that 'Trespassers will be prosecuted' are inaccurate in that most trespassers can only be sued.

Trespass in a residential property with an intention to reside is, however, a crime, under the section 144 Legal Aid, Sentencing and Punishment of Offenders Act 2012, meaning that police can assist in eviction.

There are also specific forms of trespass which are a criminal matter (per the Criminal Justice and Public Order Act 1994, as amended by the Anti-social Behaviour Act 2003). These include:

*   raves (where there are ten or more people attending or waiting for rave to start);
*   travellers (where there are more than two people with an intention to reside);
*   trespass accompanied by intimidation of persons engaged in lawful activity (a provision which has most often been used against hunt saboteurs and other animal rights activists).

And in the interests of national security there are prohibited sites under Serious Organised Crime and Police Act 2005, as amended by section 12 of the Terrorism Act 2006, such as certain dockyards, nuclear sites, government buildings and Crown land. And trespass on railway premises has been a crime from the earliest days of the railways (Railway Regulation Act 1840, Regulation of the Railways Act 1868, British Transport Commission Act 1949).

Contract or tort – which claim should be made?

A given scenario might give rise to several causes of action, e.g. a claim in both negligence and nuisance, due to the different requirements in establishing each wrong.

Where there is an apparent lack of care, then if there is a contract between the parties, that will normally be the first cause of action. But lack of care might also give rise to a negligence claim.

The key practical reason for claiming in tort where a contract exists is that an action in contract must be brought within 6 years of the formation of the contract (Limitation Act 1980). In tort (other than for personal injury, which has a limitation period of 3 years), the limitation is 6 years from when the wrong is apparent, which could well be some time later (with an absolute limitation of 15 years). So if out of time in contract, then tort might still be available.

Where there is no contract, there will be recourse to tort (subject to fulfilling the requirements).

The concurrent duty in contract and tort is a complex area and subject to considerable disagreement, the theory of which is beyond the scope of this book, but this note highlights the key points.

11.2 Crime

11.2.1 Health and safety

Whilst most areas of law encountered in everyday professional practice (such as contract, land and employment law) are civil matters, a breach of health and safety law is criminal. The consequences of this are that those accused of breaches of health and safety law are prosecuted in the criminal courts (Magistrates' Court or Crown Court) rather than sued in the civil courts (County Court or High Court). If convicted, there will be a punishment in the form of a fine (unlimited), potential imprisonment (for up to two years) and/or an order to amend the problem in the form of an improvement order or, where there is significant immediate danger, a prohibition order forbidding all operations until the problem is remedied.

Health and Safety – HSE website

The HSE (Health and Safety Executive) website is vital reading for all working people, whether employers, employees, workers or self-employed. There is up-to-date guidance on the law and legal compliance along with practice guides and much other information:-www.hse.gov.uk.

Health and safety law is governed by the Health and Safety at Work etc. Act 1974. This Act stipulates that employers must ensure the health, safety and welfare of employees (section 2) and those other than employees (section 3) so far as is reasonably practicable. Thus the duties extend to workers, sub-contractors, visitors to premises such as customers, members of the public, i.e. anyone affected by the business and work being conducted.

The general obligations under the 1974 Act are somewhat vague and much detail has been added by subsequent regulations. The most important of these (subject to industry-specific provisions) are often referred to as the six-pack, as they were introduced together, under European directives, and have been subject to amendment since first introduced:

Management of Health and Safety at Work Regulations 1999

These regulations introduced a systematic approach to risk assessment. They also cover the requirement of a written health and safety policy and risk assessment records where there are more than five employees, and that employees have appropriate training and supervision.

The remaining regulations are as follows. Full details and free explanatory leaflets can be found on the Health and Safety Executive website.

- Health and Safety (Display Screen Equipment) Regulations 1992
- Manual Handling Operations Regulations 1992
- Personal Protective Equipment at Work Regulations 1992
- Provision and Use of Work Equipment Regulations 1998
- Workplace (Health, Safety and Welfare) Regulations 1992

There has been a lightening of the burden on self-employed persons, particularly those without employees, since 2015 (under the Deregulation Act 2015 and the Deregulation Act 2015 (Health and Safety at Work) (General Duties of Self-Employed Persons) (Consequential Amendments) Order 2015). The designation as self-employed has received much legal and media attention over recent years and is a complex area, both regarding health and safety obligations and for employment law and taxation purposes. The consequences of incorrectly ascribing self-employed status can be extremely costly.

Self-Employed Status – HMRC guidance

This area is complex and there can, indeed be different designations for health and safety and taxation, so advice should be sought. HMRC provide a good basic overview – www.gov.uk/working-for-yourself/what-counts-as-self-employed.

If, sadly, a fatality arises in a work/employment context where a breach of the law is considered, the most common action is a prosecution under the Health and Safety at Work, etc. Act. There may, however, be a prosecution for gross negligence manslaughter, where an individual can be ascribed with such lack of care that such a crime is deemed to be appropriate. This could result in a fine or up to life imprisonment.

Where a company or partnership is involved, there might, in the alternative, be a prosecution under the Corporate Manslaughter and Corporate Homicide Act 2007. This allows the offence of manslaughter to be charged when no one individual is deemed (or can be identified) to be culpable but where, rather, the death is felt to be caused by the way in which a firm's activities 'are managed or organised'. Clearly a company or partnership cannot be imprisoned, as an inanimate entity, but the offence carries a recommended sentence of no less than £500,000 and publicity orders highlighting the matter.

11.2.2 *Bribery*

The Bribery Act 2010 refers to 'a financial or other advantage' gained through the improper performance of a function or activity, i.e. work influenced wholly or in part by a bribe, rather than in good faith on normal market principles. In addition to overt payment a bribe might be in the form of gifts, corporate hospitality, promotional expenses, travel expenses and accommodation costs, vouchers or other cash equivalent and provision of services such as the use of a car or provision of a decorator.

The Act consolidates the previous common law and statutory offences dating back to the nineteenth century and replaces these with two simple general bribery offences concerned with giving (active bribery) and receiving bribes (passive bribery).

For a person to commit the active offence of giving a bribe, there has to be intention and knowledge, in addition to the giving of the bribe. Both the active and passive offences incorporate the notion of 'improper performance', or 'wrongfulness'. The key to whether an offence has been committed is the connection between the bribe and this wrongfulness element; without that connection, no offence is committed. What is 'improper' is assessed by reference to whether a reasonable person would consider the recipient of the bribe to have breached an expectation of good faith, impartiality or trust. These general offences apply to activities of a private or public nature and apply equally to both individuals and corporate entities.

Under section 11 of the Act, an individual guilty of an offence under section 1 or 2 (giving or receiving a bribe) is liable to a maximum prison sentence of ten years, a fine, or both.

The Act was criticised for being unclear in relation to the issue of corporate hospitality. However, as the Guidance makes clear, 'the Act does not aim to stop corporate hospitality *per se*, but simply to prevent bribery under the façade of corporate hospitality'. As former Director of the Serious Fraud Office (SFO), David Green QC, has been quoted as saying: 'The sort of bribery we would be investigating would not be tickets to Wimbledon or bottles of champagne. We are not the "serious champagne office".'

Bribery – key information

Detailed up-to-date information can be found on the SFO website: www.sfo.gov.uk/.

11.2.3 *Fraud*

Fraud is securing unfair or unlawful advantage through intentional deception. The law is governed by the Fraud Act 2006 which includes a number of related offences:

Section 2, false representation with an intention of gain or causing loss to another. Note that there need not be actual gain or loss, only that there is an intention of such. Also, the perpetrator must know that that statement is *or might be*, untrue or misleading. This ensures that activities like high risk investment schemes can be caught where assertions are known to be potentially misleading. It also 'catches' so-called mortgage fraud where lenders are induced to lend against incorrect or incomplete information.

Section 3, failure to disclose information when there is legal duty to do so. This element of the Act covers situations such as where there is a duty to disclose information, based on contract law or legislation, perhaps stemming from the relationship, e.g. between professional adviser and client.

Section 4, abuse of position. This covers a variety of situations where there is an obligation to protect the victim and would include the client relationship or an employee taking unfair advantage of their position.

In addition to not committing fraud, businesses must guard against being the victim of fraud. Good fraud management practice involves putting in place a 'fraud response plan' which will incorporate procedures for detection, investigation, prevention of further loss, securing evidence, recovery and reporting.

Exposure in this area involves proper systems for handling client monies. Chartered surveyors should refer to the RICS Professional Statement on *Client Money Handling*, operative from January 2020.

An increasing area of fraud related to property occurs where a fraudster sells a property without the owner's knowledge. The purchaser loses their money when the Land Registry stops the transfer, or the owner can face losing their property if the Land Registry does not detect the fraud. Property Alert is a free property monitoring service run by the Land Registry aimed at anyone who feels that a registered property could be at risk from fraud.

Fraud – Land Registry alerts

Email alerts are sent when certain activity, such as official searches, occurs on the monitored properties, allowing action where necessary. See https://propertyalert. landregistry.gov.uk/

11.2.4 Money laundering

Money laundering refers to the unlawful process of concealing the origins of money obtained illegally by passing it through a complex sequence of transactions, whether through banking or commercial activities. Such schemes aim to return monies to the launderer in an indirect way. The law is complex and is governed largely by the Proceeds of Crime Act 2002, but also by the Terrorism Act 2000, the Anti-terrorism, Crime and Security Act 2001, the Serious Organised Crime and Police Act 2005 and the Sanctions and Anti-Money Laundering Act 2018. And there is, of course, an overlap with the bribery provisions, noted above.

If your business is supervised by HMRC for anti-money laundering purposes, you will need to meet the approval requirements covered under the Money Laundering, Terrorist Financing and Transfer of Funds (Information on the Payer) Regulations 2017. Businesses required to register include estate agents. See HMRC 'Estate agency guidance for money laundering supervision' for further information.

Money laundering – RICS guidance

Guidance for all surveyors and valuers and RICS-controlled business can be found in the RICS Professional Statement, published in February 2019, 'Countering bribery, corruption, money laundering and terrorist financing'.

11.3 Types of business

[T]There are four main legal structures under which a business may be operated, and all relevant forms should be explored to ensure the best structure as there are significant differences in terms of liabilities and the tax position. There are also some other forms which are noted below for information.

Legal structure of businesses – Companies House guidance

Companies House has guidance on requirements of business registration, on-going compliance and disclosure for limited companies, limited liability partnerships and other company types: www.gov.uk/government/collections/companies-house-guidance-for-limited-companies-partnerships-and-other-company-types.

11.3.1 Sole trader

Someone operating alone is a sole trader. Whilst there is an obligation to inform Her Majesty's Revenue and Customs when trading starts, there are no other formal requirements. A sole trader can trade under a business name, subject to certain restrictions which can be found on the Companies House website, and sections 1200–1208 Companies Act 2006.

11.3.2 Partnership

If there is more than one person in the enterprise, then there is a partnership. If no formalities are gone through, then the Partnership Act 1890 has default provisions, e.g. that profits will be split equally. However, the drawing up of a partnership agreement is strongly advised to avoid either an undesired default position or disputes, and for personal protection from such mechanisms as joint and several liability (whereby each partner is personally liable for the entire partnership debt). For example, rather than running the risk of such liability for debts run up by others, the agreement could state that every transaction above a given sum requires agreement of two partners (or whatever is deemed appropriate). Partnership profits are distributed according to agreed percentages between the partners and each partner pays income tax accordingly.

It is possible, where there is at least one partner with unlimited liability (as is the usual position) for one or more partners to have limited liability under the Limited Partnerships Act 1907. These are often 'sleeping' partners, i.e. those who may have invested monies but who are not involved in the day-to-day running of the business.

11.3.2 Limited company

In a commercial context we are considering companies limited by shares. Note that there are other forms with non-commercial purpose such as companies limited by guarantee.

A private limited company is indicated by the word Limited or Ltd after the business name (or the Welsh equivalent) and means that an individual shareholder's liability for unsecured debt is limited to the extent of their unpaid share capital. Unlike sole traders and partnerships, this structure requires formal establishment with registration at Companies House. Along with registration, there is a measure of formality in the running of a company, with public information about directors, annual returns and the lodging of audited account (although in short form for small and medium-sized enterprises). Whilst in a small business, the directors and shareholders may well be the same individuals, they need not be so. The business pays Corporation Tax on profits with shareholders paying income tax on their dividends and directors paying income tax on salaries, bonuses and, where they are also shareholders, dividends.

Companies are a separate legal entity from the directors and shareholders (*Salomon v A Salomon & Co. Ltd* (1897)), unlike partners in partnership. The company can be sued or prosecuted in its own right and individuals cannot (in the absence of fraud or wrongful trading) be held liable for the company's debts. It should be noted that the benefit of limited liability can be lost where either personal guarantees are given (as is likely to be required in the case of small business loans), or where there is a suspicion of fraud, or that corporate status was established for the sole purpose of separating individual and corporate personalities (*Gilford Motor Co. Ltd v Horn* (1933)).

11.3.3 Limited liability partnership

A relatively new business form, popular with many professional businesses such as solicitors and surveyors, is the limited liability partnership (under the Limited Liability Partnerships Act 2000). The name is somewhat misleading as this is somewhat closer to corporate structure than partnership. The business must be registered with Companies House and submits annual accounts. The business is run by the members. Members are not each liable for the whole LLP debt, only jointly liable per the extent of any agreement. The profits of an LLP are taxed like partnerships, i.e. each member is taxed on their share.

11.3.4 Other business forms

Other forms of business include the **public limited company** (plc) which is only appropriate for larger organisations due to the minimum share capital requirements. A public limited company is indicated by the letters plc after its name (or the Welsh equivalent) and needs a minimum share capital of £50,000 and its shares may be traded on a stock exchange. A relatively small number of property and surveying firms operate under this form. **Co-operatives** are a popular form in some spheres, e.g. certain agricultural sectors, although less so than in mainland Europe. And there are relatively new structures for specific purposes such as **Community Interest Companies** (CIC), introduced under the Companies (Audit, Investigations and Community Enterprise) Act 2004 for operations that want to direct their profits and assets to the public good.

11.4 General Data Protection Regulation (GDPR)

The General Data Protection Regulation (GDPR) was introduced in 2016. The regime is covered under the Data Protection Act 2018 and is designed to give citizens more control over, and protection of, their personal data in the light of an increasingly digital business and personal culture.

Under the terms of GDPR, organisations have to ensure that personal data is gathered legally and under strict conditions, and those who collect and manage data are obliged to protect it from misuse and exploitation, as well as to respect the rights of data owners. There are criminal penalties for not doing so.

The types of data considered personal under the existing legislation include name, address, and photographs.

Hacking: Organisations are required to notify the appropriate national bodies as soon as possible in order to ensure appropriate measures can be taken to prevent their data from being abused.

Access: Consumers have rights to easier access to their own personal data with organisations required to detail how they use customer information in a clear and understandable way.

Mailing lists: Consumers need to agree to their details being retained, e.g. to be on a mailing list there must be tacit agreement, not simply an 'opt out'.

Failure to comply with GDPR can result in criminal charges and significant fines.

GDPR – key information

Details on compliance with GDPR can be found on the Information Commissioner's Office (ICO) website (https://ico.org.uk) and from the RICS (www.rics.org/uk/footer/gdpr/) and should be reviewed by all businesses to ensure compliance. Although recent changes in the law have gained much publicity, the ICO stresses a balanced approach of 'comprehensive but proportionate' measures.

11.5 Equality

Anti-discrimination law started tentatively in the nineteenth century as provisions to give greater rights to women were brought in (e.g. the Married Women's Property Act 1882) but the first modern equality law came with the Equal Pay Act 1970, the Sex Discrimination Act 1975 and the Race Relations Act 1976. These Acts, together with others introduced over the subsequent three decades, were consolidated and amended in the Equality Act 2010 which makes it unlawful to discriminate against anyone on the basis of nine protected characteristics:

- age
- disability
- gender re-assignment
- marriage and civil partnership
- pregnancy and maternity
- race
- religion or belief
- sex
- sexual orientation

A person may also not be discriminated against on the grounds of trade union membership or activity (Trade Union and Labour Relations (Consolidation) Act 1992).

Equality and discrimination – key information

Along with government guidance (www.gov.uk/guidance/equality-act-2010-guidance), the ACAS website is a valuable tool on all employment matters: www.acas.org.uk.

12 Valuing in the UK

The future and conclusion

12.1 Overview

In putting together this book, we have used the original *Residential Property Appraisal* as our base and adapted it to reflect changes that have taken place over the last two decades. Changes will continue to happen, but the core principles of valuation should stay the same. Which property is similar to another? What did that sell for? Was that an arm's length transaction? The resulting conclusion will give a good indication of what the market is doing. Sounds simple but each property is unique, and the nuances can cause value to vary, even more so in the higher value, more unique properties. However, that is what makes the job interesting and a key component in adding value for our customers.

We have tried to enlighten the reader that an interest in land goes beyond just the bricks and mortar and even just its condition, so much more has now become important in making buying decisions.

There was a distinct lack of guidance in how to do a residential valuation at the time the first edition of this book was published, and the situation has not improved much. The original Red Book gave some context, but in recent years there has been a move to present the Red Book as a high-level Standard and to support those Standards with guidance, some of which has been residential focused (New Build, Japanese knotweed and few others). However, the suite of available guidance is not comprehensive. This book looks to fill some of these gaps.

12.2 The influence of the courts, fraudsters and mortgages

In the 1980s, guidance was provided by the courts in identifying areas of anomaly in general practice and identifying such matters as the 'trail of suspicion' and for whom there is a duty of care, to give just a couple of key areas. The role of the Red Book following these early cases was to provide a Standard to which valuers should operate and therefore not rely on the courts telling us retrospectively what should be done. It seems inevitable that every time there is a property recession, this is followed by a spate of cases identifying the shortfalls. This happened after the recession of the early 1990s and the subsequent one in 2009–10. Hopefully, this book will identify good practice and allow valuers to argue the case for common-sense good practice, to ensure they do not succumb to those who would like to see corners cut and values inflated for their own self-interest.

DOI: 10.1201/9780367816988-12

Property is, for over 20 million people, a huge investment, and when the figures involved are usually six figures or more, there is a great incentive for the fraudsters, but it also plays a huge part in the economy of this country. Therefore, understanding how valuations work and how comparables are analysed is an important feature for any valuer. It is no surprise that those relying on the valuations expect a degree of competence and transparency in how a figure has been arrived at. It is hoped that this book has lifted some of the veil of secrecy and provided some techniques that can be used. However, it takes time to develop the skill of undertaking valuations, knowing your market, knowing what is more attractive to a buyer and knowing what has sustainable value.

There is a huge focus in the residential valuation profession on the mortgage market, because that is where the majority of funding is found to finance the acquisition of a property. Removal of that funding has a profound effect on the market. This is why any residential valuation has to have reference to the open market and its influence of available funding to assist with affordability.

12.3 The future – technology, risk and information

In recent years there has been a significant move towards automated valuations and the technology behind this is probably the key to advancing digital tools for the valuers. There are increasing amounts of data now available that can be interpreted by the valuer to give more reliable assessments on property values, particularly those where values may not be as sustainable as others. Care must, however, be taken to check the provenance and integrity of any data sources you use, as not all information on the internet is correct. We all know of properties that we would love to own and if they ever come on the market they sell very quickly in whatever market. However, these are the minority, there are many more where they are marketed and do not sell as quickly and some stick on the market. They may need a bit of love and attention or they may have a garden facing the 'wrong' way, or there could be a whole host of reasons. These nuanced reasons are all there, but mostly in the mind of the experienced valuer. Getting them in the rationale with evidence to support why they make a difference is the key to being less vulnerable to a claim going forward.

Many countries are more advanced in their data collection and how they interpret that data. This will come in the UK and it is hoped that this book will provide the basics of how to start using that data in a transparent way.

Risk is a key component to lenders and they, as clients, need to have confidence that the valuation profession understands their perception of risk and that it is reflected in the information provided to them. This is an education for both sides.

As customer standards increase and we all expect a better quality of living, then housing needs change and we now have to consider things that were an unbelievable luxury 50 years ago. In a typical property having an internal bathroom, some form of central heating and no damp were reasonable expectations. At the time of writing, a property that does not have all the modern features can often be seen as a blank canvas and it is not such a big task to get the builders in to create the perfect home. However, there are other things that change, urban living was very acceptable in the early part of the twenty-first century, but the arrival of COVID-19 in 2020 changed this and rural living has become the new trend or at least having a bigger garden, with outside facilities. Will this trend be sustained in future years?

Having a good broadband facility is as essential as having a heating system and not only are thermal efficiencies recorded but pollution levels can now also be very significant. The World Health Organisation (WHO) publishes levels of pollution that are acceptable, and most countries now have targets that have to be achieved. Therefore, will these play an important part in the factors that materially affect value in the future? The modern-day valuer needs to be aware of what information is freely available and gather evidence of how the attributes applicable to a property may influence value.

In the future, lenders and valuers need to work together more closely to ensure that emerging risks are effectively managed. Matters such as the invasive plant Japanese knotweed, or spray foam insulation have caused lenders to be concerned about the long-term impact of these problems on their securities. It is important that valuers, as the experts, look to advise their clients of these risks in a balanced way, ensuring that a problem does not get overlooked or, indeed, over-emphasised. In this way, valuers can add real value to their clients. The key to confidence in the valuation profession is to ensure that surveyors are consistent and proportionate in the advice they give to lenders. It is hoped that this book helps provide some simple guidelines as to how such matters need to be approached.

A lender needs certainty in order to proceed but, given the time frames involved in the housing transaction process, certainty cannot always be achieved in the short-term, hence the need for the valuer to make assumptions. In some cases, these are assumptions that something exists when it does not, e.g. when a buyer comes forward to buy a new property not yet constructed (known as a special assumption). In order for the transaction to proceed, a conditional offer of mortgage needs to be made. This relies on a valuation, but at the time the property may not have been completed, so the valuer has to make a 'special assumption' that the property is complete and, as part of that assumption, that it is constructed as the plans suggest and to the standard specification expected.

Where there is a lack of certainty and assumptions are made, some of which will have health and safety consequences, then all parties have to accept the risks in doing that and if they are not prepared to accept the risk, then they must undertake further investigations to establish certainty. This could be by obtaining the services of a structural engineer or awaiting the outcome of legal searches or an investigation into a lease. Cutting the corners and relying on the assumptions may not always have a satisfactory outcome but all parties need to be aware of that before exchanging the contracts.

In some cases, it has not always been possible to achieve the level of certainty required, as the assumptions have been too extreme. The following examples for high rise flats may help the understanding of this, as a conclusion of how to approach the valuation.

12.4 The example of flats

The concerns with the fire safety of cladding on flats followed a serious fire in London in 2017 in which a number of people lost their lives. As the importance of the part played by the retro fit cladding in the rapid spread of the fire became apparent, the government looked to address the issue by publishing guidance to freeholders. This required building owners to assess their properties and to remove combustible materials from the exterior. This guidance posed a particular challenge to valuers.

Where replacement of cladding is necessary, this is an expensive repair and it is likely that freeholders will pass the cost of this down to the leaseholders, with a resultant large one-off expense. Possible fire safety risks to a building and the prospect of a hefty charge in the future are likely to impact the value of the property, but without knowing the materials with which the building is clad, how can a valuer assess this value impact (or know if there is any impact at all)? In order to achieve certainty, it is necessary to identify the type of cladding and determine whether it is a fire risk. This is both an expensive and time-consuming exercise, with many arguments about who should be liable for the cost.

It is not too dissimilar to the situation when it was discovered that a number of high-rise buildings were constructed with what was known as Large Panel Systems (LPS). These panels were literally like a house of cards but designed to be securely fasted at all the edges of the panels. However, it was discovered that some had not been correctly bolted together and a building in London partially collapsed. As Wikipedia reports:

> Ronan Point was a 22-storey tower block in Canning Town in Newham, East London, which partly collapsed on 16 May 1968, only two months after it had opened. A gas explosion blew out some load-bearing walls, causing the collapse of one entire corner of the building, which killed four people and injured 17. The spectacular nature of the failure (caused by both poor design and poor construction), led to a loss of public confidence in high-rise residential buildings, and major changes in UK building regulations resulted.

Sound familiar? The cost of identifying the buildings with such systems and determining whether they were correctly constructed was huge. Provisions were made for the requirement of a structural appraisal before any block identified with this type of construction could be considered for mortgage. As this appraisal required taking a look at a sample of joints, it was invasive and, given the height of the buildings, this just added expense. Few management companies wanted to undertake such an inspection and certainly no single vendor could afford such certification. The result was that lenders would not lend. Exactly the same as with the cladding situation, possibly more lenders are committed now than back in the 1960s and 1970s, because back in the time of the Large Panel System, there were few loans on high rise property, as it was generally seen as too risky and there was a general rule of nothing acceptable over four storeys apart from within London.

In recent years high rise blocks have been treated as for any other mortgage, but in fact, the reality indicates that there are all sorts of issues that need to be considered:

- The disproportionate cost of maintenance, due to need for scaffolding or cradles for work to the main walls and windows. Work to the roof could involve a crane.
- Fire risk has been shown to be a huge issue and all sorts of regulations are now being refined, enhanced and introduced.
- The buildings are leasehold, and the management of the leaseholders and their properties needs a competency that cannot always be left to a management committee of owners, although some can work successfully. There are also some very good management companies used to run the larger estates, but this comes with a cost.

- Hazardous materials can give rise to huge expenditure found generally on older buildings, but given the nature of the building, the treatment of such has to apply to all the residents and that could be hundreds of families.
- Social order, or possibly more bluntly disorder, is a bigger challenge when literally living on top of one another and this is where the original build standard can play a big part. It should be asked whether there was sufficient fire resistance included and also appropriate thermal and noise provisions.
- Given the leasehold nature of the building, all the costs for the ongoing maintenance, repair and servicing of the common parts have to be covered by service charges and unforeseen issues can create financial burdens for leaseholders.

These are just a few of the issues that need to be considered with high rise living or flats. Undoubtedly, with the need for more housing we have seen an increase in this type of living.

Figure 12.1 is a low rise block of flats and we just want to highlight a few points that raise questions:

- It can already be seen that the design of the parapet wall has resulted in premature staining, which, unless corrected, will be an ongoing eyesore that could affect flats within the whole block. What provisions are there to remedy this?
- The balconies are constructed with timber boarding and these may need to be replaced due to potential fire risk.

Figure 12.1 Modern block of flats already showing design flaws

- The nature of the roof, although attractive, poses a challenging area in respect of the valley that has been created. Are there some provisions for access to check for any blockages?
- Some of the other blocks in this development also have cladding of a type that is a fire risk. Do any of those blocks come under the same lease?

All those questions need to be asked and outcomes delivered for a valuation to be created with any degree of certainty. These questions come immediately to mind and there may be more on inspection, many of which are not typical for a house on two floors. If the answers cannot be found, then are the assumptions as to the value going to be of a type that create an acceptable risk?

In the case of cladding generally, many loans have already been issued, but due to the lack of certainty, lenders are reluctant to process more loans and such impacted properties are facing challenges in the mortgage market and the resultant impact on value. It is not possible to create an assumption that a solution will be forthcoming, because the assumption is not realistic. The consequences of that assumption being wrong and there being an issue as a result of defective cladding could lead to the application of an incorrect value and, at worst, to loss of life. Therefore, no-one wants to encourage the buyers to proceed in something that could be catastrophic.

Understanding when and where certainty must result in delay and assumptions are too extreme is something that makes the valuer's role more challenging, but more rewarding, because it is such advice that is helpful to all the parties to the transaction.

Valuation is not black magic and explaining the rationale supporting a valuation decision for transparency reasons is what the customers should expect.

References

Legislation

Agricultural Holdings Act 1986
Agricultural Tenancies Act 1995
Anti-social Behaviour Act 2003
Anti-terrorism, Crime and Security Act 2001
Arbitration Act 1996
Bribery Act 2010
British Transport Commission Act 1949
Charities Act 2011
Coal Industry Act 1994
Commonhold and Leasehold Reform Act 2002
Companies Act 2006
Consumer Contracts (Information, Cancellation and Additional Charges) Regulations 2013
Consumer Protection from Unfair Trading Regulations 2008
Consumer Rights Act 2015
Corporate Manslaughter and Corporate Homicide Act 2007
Crime and Courts Act 2013
Criminal Justice and Public Order Act 1994
Criminal Law Act 1977
Data Protection Act 2018
Deregulation Act 2015
Deregulation Act 2015 (Health and Safety at Work) (General Duties of Self-Employed Persons) (Consequential Amendments) Order 2015
Easter Act 1928
Employment Rights Act 1996
Energy Act 2011
Environmental Protection Act 1990
Equality Act 1970
Fire Safety Act 2021
Fraud Act 2006
General Data Protection Regulation 2016
Health and Safety at Work etc. Act 1974
Health and Safety (Display Screen Equipment) Regulations 1992
Homes (Fitness for Habitation) Act 2018
Homicide Act 1957
Housing Acts 1988, 1996 and 2004
Housing Defects Act 1984
Housing Grants, Construction and Regeneration Act 1996

Land Registration Act 2002
Landlord and Tenant Act 1954, Part II
Law of Property Act 1925
Law of Property (Miscellaneous Provisions) Act 1989
Leasehold Reform Act 1967
Leasehold Reform, Housing and Urban Development Act 1993
Legal Aid, Sentencing and Punishment of Offenders Act 2012
Limitation Act 1980
Limited Liability Partnerships Act 2000
Limited Partners Act 1907
Local Democracy, Economic Development and Construction Act 2009
Management of Health and Safety at Work Regulations 1999, as amended
Management of Houses in Multiple Occupation (England) Regulations 2006
Manual Handling Operations Regulations 1992
Married Women's Property Act 1882
Money Laundering, Terrorist Financing and Transfer of Funds (Information on the Payer) Regulations 2017
Occupiers' Liability Act 1957
Offences Against the Person Act 1861
Partnership Act 1890
Party Walls, etc. Act 1996
Personal Protective Equipment at Work Regulations 1992
Petroleum Act 1998
Prescription Act 1832
Proceeds of Crime Act 2002
Property Boundaries (Dispute Resolution) Bill [HL] 2017–19 and [HL] 2019
Provision and Use of Work Equipment Regulations 1998
Race Relations Act 1976
Railway Regulation Act 1840
Regulation of the Railways Act 1868
Rent Act 1977
Rentcharges Act 1977
Sales of Goods Act 1979
Sanctions and Anti-Money Laundering Act 2018.
Serious Organised Crime and Police Act 2005
Sex Discrimination Act 1975
Supply of Goods and Services Act 1982
Terrorism Act 2000
Terrorism Act 2006
Town and Country Planning Act 1990
Town and Country Planning (General Permitted Development) Order 1995
Trade Union and Labour Relations (Consolidation) Act 1992
Trusts of Land and Appointment of Trustees Act 1996
Unfair Contract Terms Act 1977
Workplace (Health, Safety and Welfare) Regulations 1992

Cases

A G Securities v Vaughan (1988)
Aldred's case (1610)).
Attorney General v P YA Quarries Ltd (1957)

Attwood v Bovis Homes (2001)

Axa Equity and Law Home Loans Ltd v Goldsack & Freeman (1994)

Barnett v Chelsea & Kensington Hospital Management Committee (1968)

Batchelor v Marlow (2001)

Benn v Hardinge (1993)

Bere v Slades (1989)

Bernstein v Skyviews & General Ltd (1977)

Birrell v Carey (1989)

Blemain Finance Ltd v e.surv Ltd (2012)

Bolam v Friern (1957)

British Railways Board v Herrington (1972)

BPE Solicitors v Hughes-Holland (2017)

Boycott v Perrins Guy Williams (2011)

Bukton v Tounesende (1348) *The Humber Ferry case*

Bulli Coal Mining Co. v Osborne (1899)

Candler v Crane Christmas & Co. (1951)

Cann & Sons v Willson (1888)

Caparo Industries plc v Dickman and others (1990)

Corisands Investments Ltd v Druce & Co (1978)

Coventry v Lawrence (2015)

Crow v Wood (1970)

Derry v Peek (1889)

Donoghue v Stevenson (1932)

Duke of Buccleuch v IRC (1967)

Family Housing Association v James (1990)

Fitzpatrick v Sterling Housing Association Ltd (1999)

Fryer v Bunney (1982)

Gilford Motor Co. Ltd v Horn (1933)

Harris v Wyre Forest District Council (1989)

Harrogate Borough Council v. Simpson (1984)

Hart v Large (2021)

Heaven v Pender (1883)

Hedley Byrne & Co. v Heller & Partners Ltd (1963)

Herons Court, Lessees and Management Company of v Heronslea Ltd (2018)

Holland v Hodgson (1872)

Hubbard v Bank of Scotland (2014)

Hunter v Canary Wharf (1997)

Izzard v Field Palmer (1999)

K/S Lincoln v CB Richard Ellis Hotels (2010)

Kelsen v Imperial Tobacco Co. (1957)

Kettel v Bloomfold (2012)

Lace v Chandler (1944)

Le Lievre v Gould (1893)

Lessees and Management Company of Herons Court v Heronslea Ltd (2018)

Lloyd v Butler (1990)

Malone v Laskey (1907)

McAdams Homes v Robinson (2004)

Moncrieff v Jamieson (2007)

Mundy v Trustees of the Sloane Stanley Estate (2018)

Murphy v Brentwood District Council (1991)

Palmer v Bowman (2000)

Platform Funding v Bank of Scotland (2012)

Prudential Assurance v London Residuary Body (1992)
Qureshi v Liassides (1994) unreported
R v Allen (1872)
R v Earl of Northumberland (1568) aka *The Case of Mines*
Rashid and Akhtar v Sharif and Sharif (2014)
Re. Ellenborough Park (1956)
Redstone Mortgages v Countrywide Surveyors (2011)
Roberts v J Hampson & Co. (1989)
Robinson v Chief Constable of West Yorkshire Police (2018)
Ryb v Conways Chartered Surveyors (2019)
Salomon v A Salomon & Co. Ltd (1897)
Scullion v Bank of Scotland plc (2011)
Simpson v Sandford St Martin Parish Council (2018)
Smith v Eric S Bush (1990)
South Australia Asset Management Corporation v York Montague Ltd (1996)
Star Energy v Bocardo SA (2010)
Street v Mountford (1985)
Sturges v Bridgman (1879)
Tulk v Moxhay (1848)
Ultramares Corp. v Touche (1932)
Van Oord v Allseas UK (2015)
Vowles v Miller (1810)
Wakeham v Wood (1981)
Watts v Morrow (1991)
Webb Resolutions v e.surv (2012)
Wells v Kingston-upon-Hull Corporation (1875)
Wheeldon v Burrows (1879)
Wibberley v Insley (1999)
Wilson v Messrs D M Hall & Sons (2004)
Woollerton and Wilson v Richard Costain (1970)
Yianni v Edwin Evans (1981)
Young v Bristol Aeroplane Co. Ltd (1944)
Zagora Management Ltd v Zurich Insurance plc (2019)

References

Bowles, L. (2019) 'Value of UK housing stock hits record high'. London: Savills plc. Available at: www.savills.co.uk/blog/article/274512/residential-property/value-of-uk-housing-stock-hits-record-high.aspx

Coke, E. (1979) *The First Part of the Institutes of the Laws of England.* New York: Garland Publishing. First published 1628.

Collins Dictionary (2014) 12th edn. Glasgow: HarperCollins.

de Silva, C. (updated annually) *Negligent Valuation Casebook: Key Cases from the 19th to 21st Centuries.* Nantwich: Zila Press. Available from carrie@carriedesilva.co.uk

de Silva, C. (2018) 'Negligent valuation de-constructed: What is negligence at law? What are the practicalities in helping to avoid a claim?' *Journal of Building Survey, Appraisal & Valuation,* 7(3), 262–274.

de Silva, C. and Charlson, J. (eds) (2020) *Galbraith's Construction and Land Management Law for Students.* 7th edn. Abingdon: Routledge.

DETR (1998) *Combating Cowboy Builders: A Consultation Paper.* London: Department of the Environment, Transport and the Regions.

Foster, H. and Lavers, A. (1998) 'A database of negligent valuation cases and literature'. *Journal of Property Valuation and Investment*, 16(1): 87–98.

Gibbons, E.. Wilson, J., and Driscoll, J. (consulting ed.) (2010) *Leasehold Enfranchisement Explained.* London: RICS Books.

Holden, G. (1998a) 'New homebuyer: more of your questions answered'. *Chartered Surveyors Monthly,* 7(8), May.

Holden, G. (1998b) 'New homebuyer: more questions answered'. *Chartered Surveyors Monthly,* 8(2), October.

Isaac, D. and O'Leary, J. (2013) *Property Valuation Techniques.* Basingstoke: Palgrave Macmillan.

Kelly, R. (1998) 'Designer housing scares off customers'. *The Times,* 3 October.

Law Commission (2011) *Making Land Work: Easements, Covenants and Profits à Prendre.* Law Com No. 327. London: The Stationery Office.

Mackmin, D. (1994) *Valuation and Sale of Residential Property.* London: Routledge.

Murdoch, J. (2002) *Negligence in Valuations and Surveys.* Coventry: RICS Books.

Murdoch, J. and Murrells, P. (1995) *Law of Surveys and Valuations.* London: Estates Gazette.

OSCOLA, The Oxford University Standard for the Citation of Legal Authorities (2012) Available at: www.law.ox.ac.uk/sites/files/oxlaw/oscola_4th_edn_hart_2012.pdf

Popular Housing Forum (1998) 'Kerb Appeal: The External Appearance nd Site Layout of New Houses'. BRMB International.

Rees, W. H. (1992) *Valuations Principles into Practice,* 4th edn. London: Estates Gazette.

Reville, J. (1998) 'Surveying safely – rights and responsibilities'. *Structural Survey,* 16(4): 172–176.

RICS (1991) *Surveying Safely: A Personal Commitment.* London: Royal Institution of Chartered Surveyors.

RICS (1995) *Appraisal and Valuation Manual.* London: Royal Institution of Chartered Surveyors.

RICS (1996) *Appraisal and Valuation Manual,* 2nd edn. London: Royal Institution of Chartered Surveyors.

RICS (1997) *Homebuyer Survey and Valuation.* HSV Practice Notes. London: RICS Business Services.

RICS (2014a) *Practice Guide Boundaries: Procedures for Boundary Identification, Demarcation and Dispute Resolution,* 3rd edn. London: Royal Institution of Chartered Surveyors.

RICS (2014b) *Surveyors Acting as Expert Witnesses,* 4th edn. London: Royal Institution of Chartered Surveyors.

RICS (2014c) *Surveyors Acting as Expert Witnesses, Client Guide,* 4th edn. London: Royal Institution of Chartered Surveyors.

RICS (2019a) *Valuation of Individual New-Build Homes,* 3rd edn. London: Royal Institution of Chartered Surveyors.

RICS (2019b) *Home Survey Standard.* 1st edn. London: Royal Institution of Chartered Surveyors.

Sparrow, J. *et al.* (2020) *UK Cross Sector Outlook: Residential, Commercial and Rural.* London: Savills plc. Available at: https://pdf.euro.savills.co.uk/spotlight-on/uk-cross-sector-outlook-2020.pdf.

Which (1997) *Special Report: Home Improvements.* October. London: Consumers Association.

Wilde, R. (1996) 'Making notes on a survey'. *Structural Survey,* 14(2): 8–21.

Woodley, M. (ed.) (2013) *Osborn's Concise Law Dictionary,* 12th edn. London: Sweet and Maxwell.

Wyatt, P. (2013) *Property Valuation,* 2nd edn. Chichester: Wiley-Blackwell.

Websites

e.surv Acadata House Price Index: www.acadata.co.uk/services/house-price-index/

HM Revenue and Customs: www.gov.uk/government/organisations/hm-revenue-customs

HMRC Inspectors' Manuals, VAT, Land and Property: www.gov.uk/hmrc-internal-manuals/vat-land-and-property

Office for National Statistics, particularly 'Housing – property price, private rent and household survey and census statistics, used by government and other organisations for the creation and fulfilment of housing policy in the UK': www.ons.gov.uk/peoplepopulationandcommunity/housing

UK Finance (formerly Council of Mortgage Lenders, CML): www.ukfinance.org.uk/area-of-expertise/mortgages

Index

Benchmarks index

Note: In the book we have used Benchmarks to highlight key points and a summary of these and the pages where you can find them are indexed here, before the Main index

Main index

Note: Author's note, the index does not give every reference for such words as valuation or value as they appear frequently, but initial references are quoted

Figure index

Tables

Legislation index

Cases

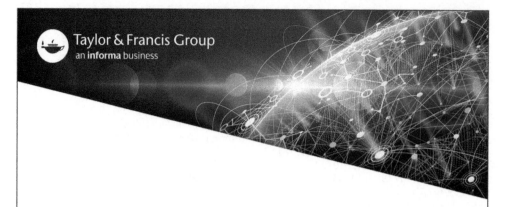